Springer Proceedings in Mathematics & Statistics

Volume 463

This book series features volumes composed of selected contributions from workshops and conferences in all areas of current research in mathematics and statistics, including data science, operations research and optimization. In addition to an overall evaluation of the interest, scientific quality, and timeliness of each proposal at the hands of the publisher, individual contributions are all refereed to the high quality standards of leading journals in the field. Thus, this series provides the research community with well-edited, authoritative reports on developments in the most exciting areas of mathematical and statistical research today.

Wenfeng Wang · Wanyang Dai · Hari M. Srivastava
Editors

Computational Statistics and Data Intelligence

APCAMS 2023, Chongqing, China, June 24–26

Springer

Editors
Wenfeng Wang
Shanghai Institute of Technology
Shanghai, China

International Academy of Visual Art
and Engineering
London, UK

Wanyang Dai
Department of Mathematics
Nanjing University
Nanjing, China

Hari M. Srivastava
Department of Mathematics and Statistics
University of Victoria
Victoria, BC, Canada

ISSN 2194-1009 ISSN 2194-1017 (electronic)
Springer Proceedings in Mathematics & Statistics
ISBN 978-981-97-4437-4 ISBN 978-981-97-4438-1 (eBook)
https://doi.org/10.1007/978-981-97-4438-1

Mathematics Subject Classification: 46N99, 46N30, 46N55, 46N10, 46N20

This Springer imprint is published by the registered company Springer Nature Singapore Pte Ltd.
The registered company address is: 152 Beach Road, #21-01/04 Gateway East, Singapore 189721, Singapore

Preface

AMS 2023

Dear Readers,
It is with great pleasure and anticipation that we welcome you to the 2023 Asia-Pacific Conference on Applied Mathematics and Statistics. As the premier forum for researchers, practitioners, and scholars in the field, this conference promises to be a platform for the exchange of ideas, discoveries, and advancements in applied mathematics and statistics.

This conference offers a comprehensive program featuring keynote speeches, plenary sessions, and technical presentations. These sessions will cover a wide range of topics, including but not limited to mathematical modeling, optimization, data analysis, probability theory, and computational methods. We are confident that this diverse program will appeal to both seasoned researchers and young scholars alike, providing them with ample opportunities to share their research findings and forge new collaborations.

We extend our heartfelt gratitude to our esteemed Organizing Committee, whose dedication and hard work have ensured the success of this conference. We would also like to express our appreciation to Technical Committee for their efforts to review the submitted manuscript of AMS 2023.

We hope that the 2023 Asia-Pacific Conference on Applied Mathematics and Statistics will prove to be an enriching experience for all attendees. May this conference inspire new ideas, foster collaborations, and pave the way for novel advancements that will contribute to the betterment of our society.

London, UK

Wenfeng Wang
AMS 2023 Conference Chair
International Academy of Visual Art
and Engineering
London, UK

Conference Organization

AMS 2023

Honorary Chair

Jinde Cao, Southeast University, China

Conference Chairs

Wenfeng Wang, International Academy of Visual Art and Engineering, UK
Wanyang Dai, Nanjing University, China
Wenwu Yu, Southeast University, China
H. M. Srivastava, University of Victoria, Canada

Program Chairs

Carlo Cattani, University of Tuscia, Italy
Jungong Han, Aberystwyth University, UK

Finance Chair

Swanirbhar Majumder, Tripura University, India

Program Committee

Muralidhar Kurni, Anantha Lakshmi Institute of Technology and Sciences, India
Alireza Valipour Baboli, Technical and Vocational University, Iran
Yoshifumi Manabe, Kogakuin University, Japan
Ahmad Fakharian, Qazvin Branch, Islamic Azad University, Iran
Vinod Kumar Verma, Sant Longowal Institute of Engineering and Technology, India
Swanirbhar Majumder, Tripura University, India
Abdelkrim Haqiq, Hassan 1st University, Morocco
Ghasem Mirjalily, Yazd University, Iran

International Technical Committee

H. M. Srivastava, University of Victoria, Canada
Ravi P. Agarwal, Texas A&M University—Kingsville, USA
Yun (Amy) Lu, Kutztown University, USA
Mingfei Qiu, Federal Home Loan Bank of Atlanta, USA
Hong-Ming Yin, Washington State University, USA
Massimo Pollifroni, University of Turin, Italy
Jozef Banas, Rzeszów University of Technology, Poland
Francisco Gallego Lupianez, University Complutense of Madrid, Spain
Carlo Cattani, University of Tuscia, Italy
Marc Paolella, University of Zurich, Switzerland
Orchidea Maria Lecian, Sapienza University of Rome, Italy
S. Joseph Antony, University of Leeds, Leeds LS2 9JT, UK
Stanislaw Migorski, Jagiellonian University, Poland
Abdalah Rababah, Jordan University of Science and Technology, Jordan
Handan Akyar, Anadolu University, Turkey
Chan Kar Tim, Universiti Putra Malaysia, Malaysia
Dang Thanh, Hue College of Industry, Vietnam
Fethi Bin Muhammad Belgacem, Paaet, Al-Ardhiya, Kuwait
Hamed Daei Kasmaei, IAUCTB, Iran
Hasibun Naher, BRAC University, Bangladesh
Haydar Akca, Abu Dhabi University, UAE
Junaid M. Shaikh, Curtin University, Malaysia
Lazim Abdullah, Universiti Malaysia, Malaysia
Mohammad Masdari, Islamic Azad University, Urmia, Iran
Moh. Ivan Azis, Hasanuddin University, Indonesia
Naoyuki Ishimura, Chuo University, Japan
Zailan Siri, Universiti Malaya, Malaysia
Tu Viet Phu, Department of Quality Management, School of Biotechnology and Food Technology, HUST, Vietnam

Contents

Method of Spare Parts Inventory Decision on the Aircraft Integrity Performance 1
Xingfang Fu

Reliability Enhancement of Solar Panels Based on the Photocurrent Equality 11
Sorena Artin

Permutation Tests for the Partial Coefficient in a Multiple Linear Regression 25
Shunpu Zhang

Research on Population Mortality Based on Generalized Partially Function Linear Model with Interaction Term 41
Kejing Mao, Songxuan Li, and Weiwei Xiao

Research on Sleep Quality Based on Generalized Partially Function Linear Model with Unknown Link Function 47
Songxuan Li, Kejing Mao, and Weiwei Xiao

Decompositions and Packing of Complete 3-Uniform Hypergraph Into Hyperstar of a Certain Type 53
Hongtao Zhao and Yuzhe Zhu

ANN Based Recognition of the Dimasa Tribal Language of North East India 65
Maidanon Langthasa, Rupesh Kumar Kanth, and Swanirbhar Majumder

Missing Value Filling Method Based on Scale Data-Correlation Coefficient Method 79
Qingcui Dong and Hong Zhao

Sixth-Order Numerical Solver Based on Truncation Error for Solution of Nonlinear Fredholm Equations 87
Chinedu Nwaigwe, Azubuike Weli, and Dang Ngoc Hoang Thanh

System Analysis and Optimization for Random Dynamic Signals 101
Xiaonan Xiao

Neighbor-Distinguishing Arc Colorings of Several Digraphs 113
Qiang Ji and Jihui Wang

Modelling Odds and Probability of Stillbirth Due to Maternal Demography (Race, Marital Status, Maternal Age and Antenatal Care) Among Women in Malaysia: Case Study in a Public Hospital at Selangor 121
Nur Syamimi Binti Haji Abu Bakar, Adriana Binti Ismail, Anita Rina Binti Ali, Nur Aqilah Alias, Norshahida Binti Shaadan, and Halilah Binti Haron

Effectiveness of Roelands Formula with Constant VPC in EHL of Higher Pair Contact 129
Jiajia Zhang, Chengzhao Liu, and Wei Ding

Numerical Simulation of Oxygen Jet Condensation in Cryogenic Liquid Rocket 141
Chengfeng Zhu and Yanzhong Li

Numerical Simulation of Mechanical Bulging Process of a Large Three-Layer Rectangular Bellow 151
Hong Jiang, De-an Meng, Kun Li, Chengcheng Zhu, Hao Zhou, and Shengdun Zhao

Effects of Different Parameter Values on the Performance of CPWC-Based Local PWV Estimation 159
Li Xiong, Yufeng Zhang, Bingbing He, Kexin Zhang, Jingying Zhu, and Xun Lang

A Robust Localization Algorithm for Intelligent and Connected Vehicles by Variational Bayesian Method 169
Ziqiang Wang, Ji Mi, Hao Zhu, and Xin Li

Method of Spare Parts Inventory Decision on the Aircraft Integrity Performance

Xingfang Fu

Abstract The aircraft integrity rate is a comprehensive indicator reflecting the performance of the aircraft integrity, and it is also a comprehensive evaluation index that reflects the quality of the objectives of the spare parts supply support system. The disadvantage of using the spare parts satisfaction rate or the expected number of stocks as a systematic evaluation indicator is that the indicator only stays in the supply system and does not go beyond the supply system to consider the impact of the supply system on the aircraft or other spare parts. Calculating aircraft uptime assesses the aircraft's ability to continue operation by combining resources over a period of time with the aircraft's combat capabilities. The aim is to determine which spare parts are prioritized for storage in order to obtain optimal aircraft uptime indicators under the condition of funding caps. But for accurate forecasting of spare parts demand, the distribution status of spare parts that generate demand must first be determined. In this paper, combined with the analysis of the statistical probability distribution characteristics of spare parts demand, taking the total cost budget constraint and the maximum spare parts satisfaction rate as the decision-making goal, this paper proposes a model for spare parts storage optimization decision that meets the optimal goal of aircraft integrity, and gives a solution algorithm of marginal analysis. The example application has verified the feasibility and practicality of the model. The model combines resources over a period of time with the aircraft's flight capabilities to assess the aircraft's ability to continue to fly in a steady-state environment. It can plan the uptime of the aircraft according to a given spare parts inventory, or it can calculate the amounts of spare parts required to complete the flight training plan task under a specified performance standard. It has solved the problem of determining which spare parts are prioritized for storage in order to obtain the maximum aircraft uptime under the constraint of limited funds.

Keywords Spare parts · Inventory decision · Aircraft integrity · Performance

---

X. Fu (✉)
College of Management, Guangdong University of Science and Technology, Dongguan, Guangdong Province, China
e-mail: xffu_11@sohu.com

W. Wang et al. (eds.), *Computational Statistics and Data Intelligence*, Springer Proceedings in Mathematics & Statistics 463,
https://doi.org/10.1007/978-981-97-4438-1_1

1 Introduction

Spare parts are an important equipment resource to ensure the integrity of aircraft, and accurate forecasting of spare parts demand has obvious military and economic benefits [1–5]. According to the requirements of equipment readiness and cost constraints, the user of spare parts collects and analyzes the supply and consumption data of spare parts and evaluates them according to the requirements of equipment readiness integrity and cost constraints, and according to the needs of use and maintenance work during the life cycle of equipment, so as to determine the quantity of subsequent spare parts. Therefore, the spare parts supply assurance system needs to meet two key objectives, one is to have a supply of spare parts whenever and wherever it is needed; The second is to minimize the cost of spare parts support on the basis of meeting military objectives [6–11]. Under the condition that there are many kinds of spare parts and limited funds and various resource conditions, if the inventory cannot be reasonably determined according to the law of demand for spare parts, it is bound to cause the inventory structure to be unreasonable, so that the limited funds cannot play the greatest benefit and reduce the spare parts support capacity. However, in the decision management of spare parts storage, the inventory optimization problem to be solved is generally some extreme value problem with constraints, and the realization of the goal is often constrained by certain conditions, such as funds, warehouse capacity, transportation capacity, management ability, etc. [12–17]. Therefore, in this paper, combined with the analysis of the probability distribution characteristics of spare parts demand, a systematic method model of spare parts storage decision is established, and the algorithm design and solution are carried out with the maximum aircraft integrity as the optimization goal [18, 19].

2 Characteristics of Statistical Probability Distribution of Spare Parts Demand

2.1 *Poisson Demand*

Order S indicates the amount of spare inventory for a spare part, the demand is the Poisson distribution of the average demand rate λ, the supplementary supply is any probability distribution $\psi(t)$, and the mean of the replenishment supply time is T, then the steady-state probability λT of the supplementary supply x spare parts is the Mean Poisson distribution [7, 8], that is:

$$h(x) = \frac{(\lambda T)^x e^{-\lambda T}}{x!} \quad 0 \le x < \infty \tag{1}$$

2.2 Compound Poisson Demand

Order s is the initial inventory of spare part, the demand is a composite Poisson process, the average demand application rate is λ, the probability of the y application for spare parts x ($y \leq x$) is expressed by revolution $f^{y^*}(x)$, then the probability distribution of the demand x is

$$p(x|\lambda) = \sum_{y=0}^{x} \frac{\lambda^y e^{-\lambda}}{y!} f^{y^*}(x) \quad 0 \leq x < \infty \tag{2}$$

If the replenishment of spare parts is an arbitrary probability distribution $\psi(t)$, and the mean of the replenishment time is T, the steady-state probability of the replenishment of spare parts x in the time period T is

$$H(x) = p(x|\lambda T) \tag{3}$$

Among them, the geometric distribution $f(x)$ is generally taken, $f(x) = (1 - \rho)\rho^{x-1} \quad x \geq 1 \quad 0 \leq \rho < 1 \, f(x) = (1 - \rho)\rho^{x-1} \quad x \geq 1 \quad 0 \leq \rho < 1$.

Yes, $f^{y^*}(x)$ is the revolution of $f(x)$, the probability distribution $f(x)$ of the sum y taken from individual samples. $\lambda = \theta(1 - \rho)$, which represents the actual average demand (demand rate) θ of spare parts in the time period T.

From the calculation formula of the mean and variance of the composite Poisson distribution, the mean X and variance V under the requirements of the composite Poisson distribution of spare parts can be obtained, respectively:

$$X = \frac{\lambda T}{1 - \rho}; \quad V = \frac{\lambda T(1 + \rho)}{(1 - \rho)^2} \tag{4}$$

2.3 Expectation of Out-of-Stocks

The number of spare parts that an entity should supply over a certain period of time but has not been able to supply from existing stocks. Using this indicator, future stock-outs can be predicted based on the demand pattern and inventory level of a spare part of the unit. Its general expression is

$$EBO_i = \sum_{X=S_i+1}^{\infty} (X - S_i)P(X|\lambda_i T_i) \tag{5}$$

Among them, EBO_i is the expected number of out-of-stocks for spare parts; S_i is the initial stock quantity of i spare parts; X is the current demand for i spare parts; $\lambda_i T$ is the mean of demand for the current period; $P(X|\lambda_i T_i)$ is the probability that the i spare parts demand X will be in the current period.

When the prior period inventory of a spare part is y_i and the actual purchase quantity for the current period is x_i, the initial inventory S_i of the spare parts for the current period is

$$S_i = x_i + y_i \tag{6}$$

1. If the current demand X_i for i spare parts of the unit follow the Poisson distribution of the parameters λ_i, There is

$$P(X_i = X) = \frac{\lambda_i^X e^{-\lambda_i}}{X!} \tag{7}$$

The expectation of out-of-stock quantity for the i spare part is calculated as follows

$$EBO_i = \sum_{X_i=S_i+1}^{\infty} (X - S_i)\frac{\lambda_i^X e^{-\lambda_i}}{X!} \tag{8}$$

2. If the demand X_i for i spare parts of the unit obeys the parameters, μ_i, σ_i, the normal distribution has a probability density

$$f(x_i = x) = \frac{1}{\sqrt{2\pi}\sigma_i} \exp\left(-\frac{(x-\mu_i)^2}{2\sigma_i^2}\right) \tag{9}$$

The expected number of out-of-stocks for this spare part is calculated as follows:

$$EBO_i = (X - S_i)\int_{S_i+\frac{1}{2}}^{\infty} \frac{1}{\sqrt{2\pi}\sigma_i} \exp\left(-\frac{(X-\mu_i)^2}{2\sigma_i^2}\right) dX \tag{10}$$

In the above two cases, the total expected number of spare parts can be added up to the expected number of n spare parts for each spare part, and the calculation formula is

$$EBO = \sum_{i=1}^{n} \sum_{X=S_i+1}^{\infty} (X - S_i)P(X_i = X) \tag{11}$$

3 Spare Parts Storage Model

3.1 Basic Assumptions

1. In a certain period, the demand for spare parts is a smooth process, such as: the support task remains unchanged, the equipment strength remains unchanged, etc.; Spare parts demand is subject to composite Poisson demand distribution;
2. All spare parts on the aircraft are field replaceable parts, and each spare part is only used in one type of aircraft and the demand cannot be met from the existing inventory, which is considered out of stock;
3. The demand for a spare part is independent of other spare parts. This assumption is true in replaceable spare parts as long as there are not many complex connections;
4. All spare parts have the same importance; Do not consider the callus of spare parts repair;
5. The time of failure of spare parts is random, and the failure is independent of each other.
6. If the spare parts fail and the demand exceeds the existing inventory, a stock-out occurs. This shortage of cargo will lead to a "hole" in the aircraft, that is, the grounding of the aircraft. Therefore, a good aircraft can be defined as: not waiting for any spare part to be repaired or in transit and cannot take off, that is, the aircraft is not obviously out of stock.

3.2 Model Building

If the number of items i representing the number of spare parts installed for a certain type of machine $i = 1, 2, \ldots j \ldots, s_i$ is the inventory of the i spare parts $s_i = 0, 1, 2, \ldots$, the expected number of i spare parts can be expressed as EBO_{i,s_i}. It is assumed that the aircraft type h consists of k h types of aircraft $h(k)$, k $= 1, 2 \ldots$ m. For the total number T_i of i spare parts installed on the h type aircraft, the probability that any aircraft with i spare parts installed may be out of stock can be expressed as $\frac{EBO_{i,s_i}}{T_i}$. Assuming that the stock is randomly distributed among all the installed parts i, the probability that the i spare parts at any location is not out of stock is $1 - \frac{EBO_{i,s_i}}{T_i}$.

Consider another situation. The percentage of counterfeit equipment parts i installed on the aircraft type $h(k)$ is $b(h(k)i)$, and if a type of aircraft does not have spare parts i, the probability value of the aircraft will not be grounded due to the shortage of spare parts $i$is 1; The probability that any other aircraft of this type equipped with spare parts i is not short of spare parts i is $\left(1 - \frac{EBO_{i,s_i}}{T_i}\right)^{Q_i}$, Assuming that other spare parts are not out of stock, the intact rate of this type of aircraft is assumed $\left(1 - \frac{EBO_{i,s_i}}{T_i}\right)^{Q_i}$.

The probability that a model of aircraft $h(k)$ is not short of spare parts is set to $q_{h(k),i,s_i}$ to indicate X_i that this type of aircraft corresponds to the good condition of spare parts i, if the aircraft is grounded due to lack of spare parts i, $X_i = 0$, otherwise $X_i = 1$. Indicates γ_i that this type of aircraft is installed with spare parts i status. If spare parts i are installed in this type of aircraft $\gamma_i = 1$, otherwise $\gamma_i = 0$, Due to the out-of-stock of fake equipment parts, they are independent of each other, The expression $q_{h(k),i,s_i}$ is

$$\begin{aligned} & q_{h(k),i,n_i} = P(X_i = 1) \\ & = P(\gamma_i = 0)P(X_i = 1|\gamma_i = 0) + P(\gamma_i = 1)P(X_i = 1|\gamma_i = 1) \\ & +P(\gamma_i = 0)P(X_i = 0|\gamma_i = 0) + P(\gamma_i = 1)P(X_i = 0|\gamma_i = 1) \\ & = (1 - b(h(k), i)) \times 1 + b(h(k), i) + \left(1 - \frac{EBO_{i,n_i}}{T_i}\right)^{Q_i} \end{aligned} \tag{12}$$

For an aircraft of a certain type, the probability that an aircraft is not short of spare parts i is calculated from the average of the probability of missing parts for each aircraft type it contains.

In this way, if the aircraft type $h(k)$ has $N(h(k))$ aircrafts, the type of aircraft h has $N(h)$ types of aircraft. Then, under the condition of the inventory n of spare parts i in the spare parts supply system, the probability that there is no shortage of spare parts i for any aircraft type h is $q_{h(k),i,s_i}$. Its formula expression is

$$q_{h,i,s_i} = \sum_{k=1}^{m} \frac{N(h(k))}{N(h)} \cdot q_{h(k),i,s_i} \tag{13}$$

Because it is assumed that the shortage process of all spare parts on the aircraft is independent of each other, and because the probability of spare parts being installed on the aircraft is different, the probability that any one aircraft h is not missing any spare parts is the product of the probability that each independent spare part is not out of stock, and we record this probability as A_h, the formula is

$$A_h = \prod_i q_{h,i,s_i} \tag{14}$$

Since a good aircraft is defined as the state of the aircraft without any lack of spare parts, it is the probability A_h that an aircraft is intact, and if considered from the perspective of the entire aircraft type, it is the percentage of the aircraft h intact for A_h, that is, the integrity rate of the aircraft h.

3.3 Optimization Algorithms

If there is a certain spare part j, the initial inventory is s_j, increase one unit of the spare parts j, and the inventory amount becomes s_j+1. Then its expected out-of-stock number $EBO_{j,s,s}$ is also reduced from the initial to EBO_{j,s_j+1}. At the same time, the probability of the aircraft not lacking a spare part j is increased from q_{h,j,s_j} to q_{h,j,s_j+1}. The intact rate of an aircraft of a certain aircraft type before the first spare part j is added

$$A = \prod_i q_{h,i,s_i} = \left[\prod_{i \neq j} q_{h,i,s_i}\right] \cdot q_{h,j,s_j} \tag{15}$$

After adding one spare part j, the aircraft's integrity rate becomes

$$A' = \left[\prod_{i \neq j} q_{h,i,s_i}\right] \cdot q_{h,j,s_j+1} \tag{16}$$

The ratio of the uptime of the old and new aircraft is

$$A'/A = q_{h,j,s_j+1}/q_{h,j,s_j} \tag{17}$$

In general form as

$$I_{h,i,s_i} = q_{h,i,s_i}/q_{h,i,s_i-1} \tag{18}$$

It can be seen that the ratio of the uptime of new and old aircraft is determined only by the inventory of spare parts. Thus, for each additional spare part, a set of data can be obtained: EBO_{i,s_i},q_{h,i,s_i} and I_{h,i,s_i}, so that the inventory of spare parts i is taken s_i,s_i+1, s_i+2, …, until the expected number of i spare parts EBO_{i,s_i} is 0. In addition, an amount is added, that is, the effect of the increase in the next spare part i per unit cost on the aircraft uptime, if the unit price of the spare part i is recorded as C_i, we define a sort value V_{h,i,n_i} for the nth spare part i, the formula expression is

$$V_{h,i,n_i} = \ln\left(I_{h,i,n_i}\right)/C_i = \ln\left(q_{h,i,n_i}/q_{h,i,n_i-1}\right)/C_i \tag{19}$$

Thus, if an aircraft initial intact rate is added $A = \prod_i q_{h,i,n_i}$, when the first spare part j (with the largest sort value) is added, the inventory of spare parts j becomes $n_j = 1$, sorted value become $V_{h,j,n_j+1} = \ln\left(q_{h,j,n_j+1}/q_{h_{h,j,n_j}}\right)/C_j$, and the new aircraft is in good condition

$$
\begin{aligned}
A' &= A \cdot \exp\left(C_j \cdot V_{h,j,n_j+1}\right) = \left[\prod_i q_{h,i,n_i}\right] \cdot \exp\left[\ln\left(q_{h,j,n_j+1}/q_{h,j,n_j}\right)\right] \\
&= \left[\prod_i q_{h,i,n_i}\right] \cdot \left(q_{h,j,n_j+1}/q_{h,j,n_j}\right) = \left[\prod_{i \neq j} q_{h,i,n_i}\right] \cdot q_{h,j,n_j} \cdot \left(q_{h,j,n_j+1}/q_{h,j,n_j}\right) \\
&= \left[\prod_{i \neq j} q_{h,i,n_i}\right] \cdot q_{h,j,n_j+1}
\end{aligned}
\tag{20}
$$

Follow the steps above to add the second spare part in turn, and the third spare part, ……, (the sorted values are the second place, the third place, ……), the additional spare parts are arranged in order of priority, written into a procurement list, and each additional spare part gets a new aircraft integrity rate, which is written in the general form as

$$
A' = A \cdot \exp\left(V_{h,j,n_j} \cdot C_j\right) \tag{21}
$$

When the first spare part is written into the list, a complete spare parts procurement list and a set of optimization schemes are obtained. Spare parts are stored sequentially in the order in which they are listed until the target specified for aircraft uptime optimization is met or the budgeted funds are equal to or just exceed the specified funds.

3.4 Examples

An application example is given here. The initial aircraft integrity rate of a certain aircraft is 0.75, due to the shortage of equipment, four types of spare parts (A, B, C, D) need to be purchased, the initial parameters are shown in Table 1, and the previous inventory is (2, 1, 1, 1), using the above optimization algorithm, you can get a spare parts procurement list. When the designated fund is 50,000, the first six spare parts (2 A, 2 B, 1 C, 1 D) can be purchased sequentially in the order listed in the procurement list, so as to obtain an optimized spare parts storage strategy (shown in Table 2). Similarly, when an aircraft uptime target of 0.91 is pre-determined, an optimized spare parts storage strategy can also be obtained according to the procurement list.

Table 1 Initial data of spare parts statistics

i	C_i	X	V	T_i	$\psi(t)$
A	3280.00	3.00	3.00	2	Poisson demand
B	4580.00	0.50	0.50	1	*Poisson* demand
C	12,680.00	2.00	2.00	1	*Poisson* demand
D	22,000.00	7.00	1	2	*Compound Poisson demand*

Table 2 Spare parts purchase list

Purchase order	c_i	$\sum c_i$	EBO^*_{i,n_i}	A_h	I_{h,i,n_i}	V_{h,i,n_i}
A	3280	3280	0.781	0.76	1.0133333	0.0000588
B	4580	7860	2.049	0.78	1.0263158	0.0000452
C	12680	20540	0.541	0.82	1.0511282	0.0000412
B	4580	42540	1.249	0.89	1.0853659	0.0000383
D	22000	47120	0.672	0.9	1.011236	0.0000254
A	3280	50400	0.41	0.91	1.0111111	0.0000243
...	...	...	...	...	...	...

4 Conclusion

The constructed spare parts storage optimization decision model takes into account not only the expected number of spare parts, but also the possible impact of spare parts on a single machine. The model combines resources over a period of time with the aircraft's flight capabilities to assess the aircraft's ability to continue to fly in a steady-state environment. It can plan the uptime of an aircraft at a given spare parts inventory, or it can calculate the amounts of spare parts needed to complete an Air Force training program mission under a specified confidence standard. Thus, in the case of a limited budget or under specified aircraft uptime targets, the model could provide a procurement list to determine which spare parts to store first in order to achieve the expected uptime targets under the most cost-effective conditions. The calculation of the relationship between this funding provided by the model and the continuity of the aircraft makes it easy for spare parts planning decision makers to identify and estimate the various types of funding required for combat spare parts.

In wartime dynamic environments, support performance indicators usually no longer consider aircraft integrity on a given day, but rather consider the guarantee supply indicators for critical days. In addition, due to the variability of repair and resupply times, the optimal combination of spare parts calculated in the pursuit of maximum uptime varies from day to day. In this way, the supply performance indicators of the security system should be continuous, that is, the aircraft integrity rate that lasts for two or more days, but the calculation process will become more complicated, because the flight plan, demand process, delivery time, repair time, and probability of not repairing in the wartime dynamic environment are dynamically changing with time, and the warehouse resupply and the shortage process are non-stationary random processes.

References

1. Rustenburg, W.D., van Houtum, G.J., Zijm, W.H.M.: Spare parts management at complex technology-based organizations: an agenda for research. Int. J. Product. Econ. **71**,177–193 (2001)
2. Hu, Y.F., Song, B.F., Wang, X.: Ascertaining of spares requirements for aircraft battle damage. Acta Aeronautica Et Astronautica Sinica **30** (3), 450–455 (2009)
3. Zou, D.X., Wang, G.N., Gao, L.Q. and Wu, J.H.: Application of improved PSO algorithm to reliability problems. J. Northeastern Univ. (Nat. Sci.) **31**(9), 1234–1237 (2010)
4. Ni, X.C., Zuo, H.F., Chen, F.T., Rong, X.: Civil aircraft rotable spare part forecasting. J. Nanjin Univ. Aeronaut. Astronaut. **41**(2), 253–256 (2009)
5. Tianke, L., Wei, L., Weixiao, S., Shicai, Y.: Reliability analysis and demand forecast of spare parts.J. Ordnance Equip. Engin. 37(4), 47–50, 80 (2016)
6. Zixu, K., Dong, Z., Huixin, L.I., Ziyue, G.: Dynamic planning technology of civil aircraft spares based on aircraft state. J. Beijing Univ. Aeronaut. Astronaut. (3): pp 1–16(2022)
7. Chen, D., Fang, Z.G., Liu, S.F.: Grey BD predication model for spare parts of repairable queuing system. Syst. Engin. Theory Pract. **40**(5), 1326–1338 (2020)
8. Kim, J., Shin, K., Yu, H.: Optimal algorithm to determine the spare inventory level for a repairable-item inventory system. Comput. Operat. Res. **23**, 289–297 (2016)
9. Yang, C., Bing, H.: Supportability grey comprehensive evaluation of radar equipment based on improved AHM-CRITIC. Telecommun. Technol. **59**(2), 229–236 (2019)
10. Military standardization center of technology and industry for national defense committee. GJB1364 implementation guidelines for cost effectiveness analysis for material. 1994:105–140
11. Lee, H.L., Moinzadeh, K.: A repairable item inventory system with diagnostic and repair service. Europ. J. Operat. Res. **40**, 210–221
12. Mabini, M.C, Gelders, L.F.: Repairable item inventory systems: a literature review. Belgian J. Operat. Res., Stat., Comput. Sci. **30**, 58–69
13. Yuan, F.S., Cui, C.L., Zhu, H.T.: Optimization model of aviation parts consumables inventory based on aircraft grounding rate. Logist. Technol. **40**(7), 132–135 (2021)
14. Peng, Z., Ting, G.: Research on inventory management strategy of aircraft parts based on system dynamics. Measur. Control Technol. **32**(11), 135–138, 143 (2013)
15. Zixu, K., Dong, Z., Huixin, L., Ziyue, G.: Dynamic planning technology of civil aircraft spares based on aircraft state. J. Beijing Univ. Aeronaut. Astronaut. (3), 1–16 (2022)
16. Jinsong, Z., Yu, H., Jun, et al. M.: Prediction method of spare parts demand based on grey model. J. Milit. Transp. Coll. **18**(1), 35–38 (2016)
17. Keviny, K.N.G., Sancho, N.G.F.: A hybrid 'dynamic programming /depth –first search 'algorithm , with an application to redundancy allocation. IIE Trans. **33**, 1047–1058 (2001)
18. Hwang, K.H., Lee, J.M., Hwang, Y.: A new machine condition monitoring method based on likelihood change of a stochastic model. Mech. Syst. Signal Process. **41**, 357–365 (2013)
19. Xiao, Z., Ye, S.J., Zhong, B.: BP neural network with rough set for short term load forecasting. Expert Syst. Appl. **36**, 273–279 (2009)

Reliability Enhancement of Solar Panels Based on the Photocurrent Equality

Sorena Artin

Abstract The world currently faces the climate change, as a serious environmental issue, which is likely to have major negative subsequence. To encounter this problem, it is supposed that renewable energy systems (RESs) should be one of the top priorities to ensure proper actions are being taken to tackle the global warming problems. As a particular case, solar cells are increasingly being installed around the world either in large and small scales to help environment and eventually reduce the greenhouse gas emission. This all means that design of the solar cells needs to be improved in order to enhance their performance and this improvement cannot be done without considering safety and reliability of the solar cells. In this talk, a solar cell with a photocurrent and a diode is considered to be investigated in terms of reliability analysis issues. For this purpose, the photocurrent equality is used to introduce a new system performance function for the solar cells. This performance function is then employed to set up a reliability analysis problem that aims at investigating the solar cell reliability. A numerical experiment is also provided to represent how this model can be used to enhance the solar cell design. Two of the most stable and efficient existing reliability analysis methods are selected to solve the proposed problems.

Keywords Renewable energy systems · Reliability analysis · Failure surface · Solar panels

1 Introduction

It is estimated that the global temperature increases by 2 °C approximately by 2050 while oil and natural gas consumption will be doubled at the same time [1]. This shows how important and challenging the next couple of decades will be to change systems' design and make them more environmentally friendly.

S. Artin (✉)
Prince Mohammad Bin Fahd University, Eastern Province 34754, Al Khobar, Saudi Arabia
e-mail: sartin@pmu.edu.sa

W. Wang et al. (eds.), *Computational Statistics and Data Intelligence*, Springer Proceedings in Mathematics & Statistics 463,
https://doi.org/10.1007/978-981-97-4438-1_2

That is why a huge number of research and industrial projects have been dedicated to developing RESs in the previous years and also have led to more investment on design and installation of systems such as solar panels and wind turbines in residential and industrial areas [2]. In Australia, as an example, it can be seen that not only sustainability factors are considered in research projects at national and regional levels, individuals are also getting more interested in adopting sustainable techniques to help environment and reduce the green gas emission [3–5].

The need and tendency to reducing the use of fossil fuels have resulted in an approach by individuals as well as governments to move towards RESs. All these issues have resulted in huge efforts to designing new systems or improving the existing ones to control and ultimately decrease the causes of climate change. However, there are different challenges involved in this process. For instance, RESs can only work upon availability of their required natural resources, such as the irradiation and the wind, which both have a random behavior. Wind farms need to be constructed where constant wind are expected.

Similarly, availability of irradiance is a significant factor for consideration when a solar panel station is planned to be installed. These all mean that environmental factors play undeniable roles in generating solar energy as radiation presence in each location affects the energy that can potentially be generated. Efficiency improvement of solar cells is also studied by investigating the open-circuit voltage and the short-circuit current density and by considering the balance between them. In this case, implementing novel nonfullerne acceptors (NFAs) has resulted in an optimal power conversion efficiency (PCE) [6].

Challenges caused by this random nature has however resulted in employing novel ideas in research activities to design RESs in more efficient ways. In this case, different data mining and machine learning algorithms are applied to improve RESs' design [7]. It has been reported that accuracy of real-time data can enhance performance of power distribution systems [8]. Moreover, to monitor solar cells functionality in large-scale solar panel stations, the Internet of Things (IoT) is employed to analyze PV cells operations [9]. Parameter estimation is also used in optimization models of solar panels where stochastic fractal search algorithms are employed to estimating the current–voltage parameters [10].

As a promising improvement, costs of power generation by RESs has already started to reduce. For instance, the cost of power generated by solar panels reduced by almost 75% between 2010 and 2017. For this purpose, the software HOMER and the particle swarm optimization (PSO) algorithms have significantly been used in the last couple of years to optimize RESs size and configuration. It is also reported that a hybrid renewable energy system (HRES) is in its optimal configuration when solar PV, wind turbine, diesel generator and battery storage are all combined. This optimum system can supply continuous power with high efficiency and low maintenance cost while its complex layout brings in high installation cost and complicated control operation [1, 11, 12]. Another method is to improve PV cells' reliability by modifying the circuit. For this purpose, a bypass diode is introduced to the circuit to decrease effects of any potential shading that could negatively impact the PV cell and the power being generated. Also, several strategies are introduced to mitigate hot spots in order to further enhance reliability of the system [13].

Employing offshore renewable energy (ORE) is another desired method to reduce fossil fuel consumption to generating electricity. Although parts of this technology, such as ocean thermal energy conversion, is not yet ready to be commercialized, it has been reported that offshore wind, wave and tidal energy technologies are available to be used in commercial scale. However, the ORE technologies are still too expensive and so not yet competitive even with other renewable resources like solar PV cells [12].

This paper aims at formulating a reliability analysis problem, as a new optimization model, for solar cells in which the photocurrent equation is used to set up a reliability analysis problem. Two variables of the photocurrent equation are considered as random variables that will then be used to formulate the objective function and constraint(s) of the problem.

2 Solar Panel Power Generation

A solar PV cell is indeed a $p-n$ semi-conductor junction where characteristic curves depend on irradiance and temperature. A commonly accepted model of a PV cell is designed with only a photocurrent resource and a diode [14, 15].

The power generated by a PV cell can be found by the basic relationship $P = I_{PV}V_{PV}$, and hence the Maximum Power Point (MPP) would be equal to $P_{max}(G, T) = \max I_{PV}V_{PV}$.

V_{PV} is the voltage of a PV cell, which is supposed to be equal to V_d in an ideal situation where V_d is the voltage of the diode of the PV cell. Also, the current of a PV cell is calculated as $I_{PV} = I_{Ph} - I_d$ where I_{Ph} is the photocurrent and I_d is the current of the diode.

Photocurrent I_{Ph} is a function of the random variables, irradiance G and temperature T as below:

$$I_{Ph}(G, T) = [I_{scn} + K_i(T - T_n)]\frac{G}{G_n} \tag{1}$$

where three nominal values in this equation are nominal short-circuit current I_{scn}, nominal temperature T_n and nominal irradiance G_n. Also, K_i is a current temperature coefficient [16].

Also, the current of the diode is formulated with respect to the temperature T and the voltage of the diode V_d as follows:

$$I_d(T, V_d) = I_S(T)\left(e^{\frac{V_d}{nV_t(T)}} - 1\right) \tag{2}$$

where n is the diode ideality factor, V_t is the thermal voltage of the junction, and I_S is the saturation current that can be found by:

$$I_S(T) = \frac{I_{scn} + K_i(T - T_n)}{e^{\frac{V_{ocn}+K_V(T-T_n)}{nV_t(T)}} - 1} \tag{3}$$

where K_V is the voltage temperature coefficient, and V_{ocn} is the nominal open-circuit voltage.

Further,

$$V_t(T) = \frac{kT}{q}N_S \tag{4}$$

where $k = 1.38 \times 10^{-23}\frac{J}{K}$, which is the Boltzmann constant, q is the electron charge, and N_S is the number of PV cells in the serial configuration.

Although there are two random variables, namely as the irradiance G and temperature T, in this relationship, reliability factors are not considered in the model. In a similar scenario however, reliability analysis problems are introduced for electricity power networks. A popular technique for this purpose is to introduce a new probabilistic constraint for a system in order to set up a reliability-based design optimization model [17].

A probabilistic constraint is proposed in the literature for electricity power networks and is then employed to ensure reliability of a power system meets the required conditions when optimizing the system [18]. It has been reported that reliability-based design optimization (RBDO) techniques can be applied to reduce the cost while also improving reliability of the system [12]. Probability considerations are also studied in multiagent networks where constrained consensus is applied. In this method, a comfortable range is proposed by each agent and then several conditions are defined based on the robustness analysis in order to guarantee convergence of an introduced method. It must further be mentioned that both fixed connectivity and time-varying connectivity are studied to ensure their potential effects are considered when analyzing the model by numerical simulations [19].

A reliability model is also introduced for the photocurrent model to maximize average of P_{max} while minimizing its variation in order to enhance system reliability. In this model, I_{scn} and K_i are considered to be bounded by lower and upper terminals [16].

Two reliability analysis problems and the most stable and efficient methods to solve them will briefly be reviewed in the next section.

3 First-Order Reliability Analysis; Problems and Methods

Reliability analysis problems are being widely employed to take safety related issues into account when designing a system. For this purpose, a range of first- and second-order problems could be applied, however, due to complexity of the second-order reliability analysis problems, the first-order problems are applied in many cases. There are two general categories of the first-order reliability analysis problems as direct and inverse problems [20] that are briefly discussed in this Section.

These problems are formulated as a constrained minimization problem using a system performance function $G(x_1, x_2, \ldots, x_n)$. Given the highly non-linear nature of these problems, the first-order reliability method (FORM) is often employed to standard normalize these problems by transferring them from original random space, i.e. X-space, to standard normal random space, i.e. U-space [21]. This transformation is done to reduce non-linearity of the problem. So, the optimization problem is set up once the problem is standard normalized and the transformation is done.

The first-order direct reliability analysis problem aims at finding the nearest point on the standard normalized failure surface $G_U(u_1, u_2, \ldots, u_n) = 0$ to the origin of the U-space. So, this problem is set up as below:

$$\begin{aligned} &\min(u_1, u_2, \ldots, u_n) \\ &s.t. G_U(u_1, u_2, \ldots, u_n) = 0 \end{aligned} \tag{5}$$

It has been reported that the conjugate gradient direction-based (CGDB) method is the most stable and efficient method to solve this problem. The optimum point of this problem is then called the most probable failure point (MPFP) [21].

While the above problem is solved to find a reliability index, which is equal to the distance from the origin of the U-space to the optimum point MPFP, a first-order inverse reliability analysis problem employs a (target) reliability index to find a point on a circular constraint with the minimum performance and so its optimum point is called the minimum performance target point (MPTP). The first-order inverse reliability analysis problem is formulated as following:

$$\begin{aligned} &\min G_U(u_1, u_2, \ldots, u_n) \\ &s.t.(u_1, u_2, \ldots, u_n) = \beta \end{aligned} \tag{6}$$

where β is the (target) reliability index.

The conjugate gradient analysis (CGA) method is reportedly the most stable and efficient method to solve a first-order inverse reliability analysis problem [20].

4 Reliability Analysis Formulation of Solar Panels

As discussed in the previous Sections, there is not any reliability analysis model available for solar cells based on a photocurrent and a diode system. However, optimization problems of the solar cells in general and reliability analysis problems of these systems in particular play undeniable role in designing RESs. That is why a new reliability analysis problem is introduced in this paper for solar cells. This model requires at least two random variables which are selected based on the existing literature.

Given the literature reviewed in the previous Sections, it can be seen that the power generated by a PV cell is a function of I_{PV} and V_{PV}, where these terms are respectively the current and voltage of the PV cell.

In this formulation, I_{PV} is itself a function of the photocurrent I_{Ph} and the current of the diode I_d while V_{PV} is ideally equal to the voltage of the diode V_d.

Given I_{Ph} is a function of G and T, and also I_d is a function of V_d and T, it can be seen that the generated power is a function of G, T and V_d among which the irradiance G and temperature T are the only random variables.

Therefore, these two random variables are considered in this paper to formulate a performance function for the solar cells that will then be used to set up the reliability analysis problems.

The coefficients and the nominal values of the photocurrent model, which are all positive numbers, are also employed to define two constants for the proposed performance function. So, based on the photocurrent function discussed in the previous sections, the new coefficients are defined as follows:

$$c_1 = \frac{I_{scn} - K_i + T_n}{G_n} \tag{7}$$

and

$$c_2 = \frac{-K_i}{G_n} \tag{8}$$

As all these nominal values and the coefficient are positive, it can be concluded that the coefficient c_2 is a negative number. Also, based on the estimated values and thresholds of these constants, it can be seen that c_2 is a very small number (in some cases, approximately zero). Also, the random variables the irradiance G and temperature T will be shown by new variables x_1 and x_2.

Therefore, the performance function is defined as below:

$$G(x_1, x_2) = c_1 x_1 + c_2 x_1 x_2 \tag{9}$$

where x_1 and x_2 are the random variables of the problem. That is, x_1 represents the irradiance G and x_2 represents the temperature T.

Based on the fundamental properties of the irradiance and temperature, the random variables x_1 and x_2 can only take non-negative values that should satisfy the following conditions:

$$0 \leq x_1 \leq G_n \quad \& \quad 0 \leq x_2 \leq T_n$$

In some cases where c_2 is almost zero, the non-linear term does not affect the behavior of the system performance function significantly, and so the existing methods would be able to solve the introduced reliability analysis problems stably and efficiently.

As discussed in the previous section, the random variables x_1 and x_2 need to first be standard normalized and then the performance function $G(x_1, x_2)$ will be converted to its standard normal version $G(u_1, u_2)$.

So, the reliability analysis problems can now be formulated using the introduced performance function (9). Therefore, the first-order direct reliability analysis problem is set up in the U-space as below:

$$\begin{aligned} &\min ||(u_1, u_2, \ldots, u_n)|| \\ &s.t. c_1(\sigma_1 u_1 + \mu_1) + c_2(\sigma_1 u_1 + \mu_1)(\sigma_2 u_2 + \mu_2) = 0 \end{aligned} \tag{10}$$

where u_1 and u_2 are the standard normalized random variables and the constraint is found by using the standard normalized performance function $G(u_1, u_2)$.

Also, the first-order inverse reliability analysis problem is modelled in the standard normal random space as follows:

$$\begin{aligned} &\min c_1(\sigma_1 u_1 + \mu_1) + c_2(\sigma_1 u_1 + \mu_1)(\sigma_2 u_2 + \mu_2) \\ &s.t. ||(u_1, u_2, \ldots, u_n)|| = \beta \end{aligned} \tag{11}$$

where β is the target reliability index.

These problems are then to be solved by the CGDB and CGA methods, respectively, in order to find the MPFP and the MPTP. The MPFP represents a point where the solar panel is likely to stop working and fail. The MPTP also shows a point where we can expect to see the minimum performance of the system.

However, the performance function can still be modified to include more nonlinear terms that may also result in less efficiency of the reliability analysis methods to solve the highly nonlinear problems.

Moreover, it should be noted that the above reliability analysis problems can be applied to formulate a RBDO problem to enhance the system reliability while minimizing the cost.

Next section provides a numerical experiment to demonstrate the model introduced in this paper.

5 Discussion—A Numerical Experiment

A new performance function for solar panels is introduced in this paper based on the photocurrent model available in the literature. There are four different parameters in this system performance function as nominal short-circuit current I_{scn}, current temperature coefficient K_i, nominal irradiance G_n and nominal temperature T_n. The introduced reliability analysis problems are solved by the most stable and efficient methods available in the literature to solve reliability analysis problems, and given their better performance, other available methods are not applied any more. Several references are cited in the previous sections that compare performance of these methods. Table 1 shows the values of these parameters that are adopted from the literature.

The above parameters are employed when calculating the coefficients c_1 and c_2 (discussed in the previous sections). Therefore, we would find $c_1 = 3.35$ and $c_2 = -0.03$.

To formulate a new reliability analysis problem of the solar panels, let's assume random variables x_1 and x_2 represent the irradiance G and temperature T, respectively. So, the system performance function (based on the photocurrent equation) would be as below:

$$G(x_1, x_2) = 3.35x_1 - 0.03x_1x_2$$

Assuming the random variables x_1 and x_2 follow the Gaussian distribution with parameters as $x_1 \sim (14, 4.6)$ and $x_2 \sim (11, 2.4)$, the standard normalized performance function can be found as below:

$$G(u_1, u_2) = 13.892u_1 - 1.008u_2 - 0.3312u_1u_2 + 42.28$$

In the following Subsections, a direct and an inverse first-order reliability analysis problems of the solar cells are set up based on the above performance function and solved by the CGDB and CGA methods to find their optimum point.

Table 1 Parameters and their nominated values

Parameter	Short circuit current I_{scn}	Current temperature coefficient K_i	Nominal irradiance G_n	Nominal temperature T_n
Value	$4A$	$0.3\frac{A}{K}$	$100\frac{W}{m^2}$	$29.8K$

6 First-Order Direct Reliability Analysis Problem

Based on the details illustrated in this Section, the first-order direct reliability analysis problem of the solar PV cell is formulated as below:
$\min(u_1, u_2))$

$$\min ||(u_1, u_2)|| \\ s.t. 13.892u_1 - 1.008u_2 - 0.3312u_1u_2 + 42.28 = 0 \tag{12}$$

This problem is solved by the CGDB method, which is the most stable and efficient method to solve the problem. This method converges and is able to find the optimum point, i.e. the MPFP. It should be mentioned that the MPFP is the closest point on the standard normalized failure surface to the origin of the U-space. The last five design points, including the optimum point, are displayed in the Fig. 1.

The MPFP found by the CGDB method is $(x_1^*, x_2^*) = (14.0029, 111.6672)$ where $G(x_1, x_2) = 0$. The latter equality indicates that the optimum point is on the failure surface.

In other words, the result shows that the circuit is most likely to stop working and fail at irradiance $G = 14.0029$ and temperature $T = 111.6672$. It can be seen that the optimum value of both variables are within the required boundaries $0 \leq x_1 < G_n = 1000$ and $0 \leq x_2 < T_n = 298$ and so meet the conditions of the problem.

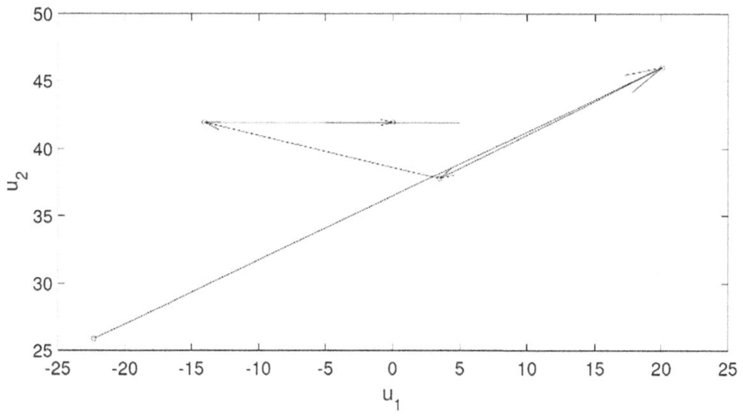

Fig. 1 Direct reliability analysis problem

7 First-Order Inverse Reliability Analysis Problem

Similar to the previous Subsection and based on the first-order inverse reliability analysis problem (6), the following problem needs to be solved to fine the MPTP:

$$\min\ 13.892u_1 - 1.008u_2 - 0.3312u_1u_2 + 42.28$$
$$s.t.\|(u_1, u_2)\| = 3 \tag{13}$$

In this problem, the target reliability index β is considered to be equal to 3.

The CGA method, which is reported to be the most stable and efficient reliability analysis method to solve this problem, converges to the optimum point MPTP $\left(x_1^*, x_2^*\right) = (0.2028, 11.1126)$ where $G(x_1, x_2) = 0.6117$.

Figure 2 represents the circular constraint of the problem and the last five design points. As these points are close to each other, a zoomed-in situation is also shown in the Fig. 3 where it can be seen that the optimum point lies on the constraint and so meets the requirements.

Therefore, based on the optimum point found by the CGA method and its relevant system performance function value, it is expected that the solar panel has the minimum performance at irradiance $G = 0.2028$ and temperature $T = 11.1126$. It is also observed that both optimum values meet the conditions of the random variables as $0 \le x_1 < G_n = 1000$ and $0 \le x_2 < T_n = 298$ in this problem.

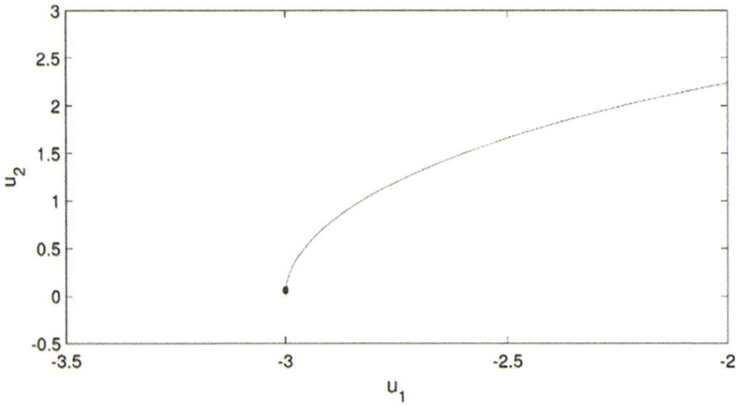

Fig. 2 Inverse reliability analysis problem

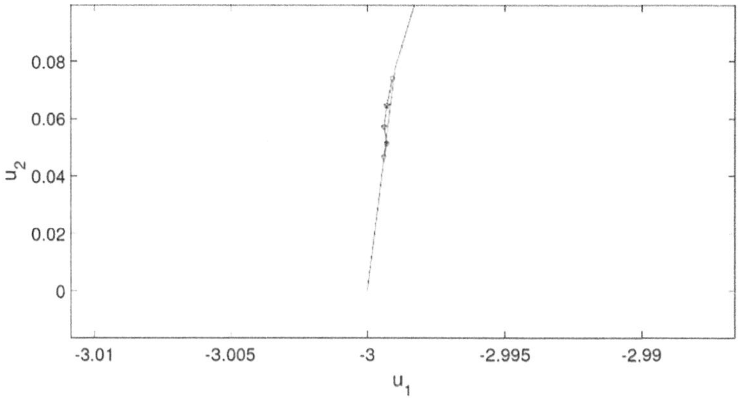

Fig. 3 Inverse reliability analysis problem (zoomed-in)

8 Conclusion and Future Works

RESs are currently the preferable resource to move away from fossil fuels. However, there are still several challenges related to the RESs that need to be addressed. One of the most important aspects is to consider reliability related issues of these systems.

In this paper, a new performance function is introduced for solar cells and is then applied to set up first-order reliability analysis problems. These problems can be solved to find the optimum points where the system is likely to fail or is expected to have the minimum performance. The most stable and efficient reliability analysis methods are employed to solve the problems.

The results demonstrated in this paper show at which point the system may fail, and so there needs to be alternative electricity power resources to supply the power. For instance, as shown in the numerical experiment, a system under the conditions discussed and assumer in this paper is very likely to stop working once the temperature reached 111. This indicates that a second energy resource as a plan B is required when designing this system.

When designing a first-order reliability analysis problem based on the photocurrent equation and using the data available in the literature, it is found that the irradiance G does not move much to get to the failure mode, while the temperature T reaches to its optimum point of the problem, which is indeed the failure point of the system, when it is far from the initial value.

On the other hand, the introduced inverse reliability analysis problem of solar cells indicates that the irradiance G gets too close to zero when the system fails, while the temperature T does not move far away from its initial value when the system turns to failure mode. It should also be mentioned that as an inverse reliability analysis problem is solved in this case, the target reliability index is supposed to be equal to a specific value (which is 3 in this paper). This constant is indeed used to make the circular constraint of the problem.

These findings can also be used to model a RBDO problem for solar cells where the cost is minimized while the system reliability is taken into consideration as well to ensure conditions on a threshold of the system reliability is met. In this case, the introduced first-order reliability analysis problems will be used in the inner loop of a bi-level RBDO problem. Furthermore, these reliability analysis problems can be studied in the future works by other scholars in a reliability analysis model of HRESs in which solar cells and wind turbines and sometimes diesel generator are combined to ensure high performance of the system.

References

1. Al-Shahri, O.A., Ismail, F.B., Hannan, M.A., Lipu, M.H., Al-Shetwi, A.Q., Begum, R.A., Al-Muhsen, N.F., Soujeri, E.:Solar photovoltaic energy optimization methods, challenges and issues: a comprehensive review. J. Cleaner Product. **284**, 125465 (2021)
2. Zhang, B., Wang, M., Su, W.: Reliability analysis of power systems integrated with high-penetration of power converters. IEEE Trans. Power Syst. **36**(3), 1998–2009 (2021)
3. Artin, P., Artin, S.: Bank of Sustainability: A business approach to motivate people to engage with sustainability. Independ. J. Manage. Product. **10**(3), 989–999 (2019)
4. Arin, P.: Critical sustainability factors of regional SMEs; a case study of regional Australia. Current Res. Environ. Sustain. **4**, 100138 (2022)
5. Nine News Australia, Neighbourhood Dispute Erupts Over Man's Backyard Wind Turbine (2022)
6. Hai, J., Song, Y., Li, L., Liu, X., Shi, X., Huang, Z., Qian, G., Lu, Z., Yu, J., Hu, H., Chen, S.: High-efficiency organic solar cells enabled by Chalcogen containing branched chain engineering: balancing short-circuit current and open-circuit voltage, enhancing fill factor. Adv. Funct. Mater. 2213429 (2023)
7. Vennila, C., Titus, A., Sudha, T.S., Sreenivasulu, U., Reddy, N.P.R., Jamal, K., Lakshmaiah, D., Jagadeesh, P., Belay, A.: Forecasting solar energy production using machine learning. Int. J. Photoenergy 7797488 (2022)
8. Gooding, P., Makram, E., Hadidi, R.: Reliability analysis of distributed generation for island scenarios utilizing Carolinas data. Electric Power Syst. Res. **107**, 125–132 (2014)
9. Shakya, S.: A self monitoring and analysing system for solar power station using IoT and data mining algorithms. J. Soft Comput. Paradigm **3**(2), 96–109 (2021)
10. Rezk, H., Babu, T., Al-Dhaifallah, M., Ziedan, H.A.: A robust parameter estimation approach based on stochastic fractal search optimization algorithm applied to solar PV parameters. Energy Rep. **7**, 620–640 (2021)
11. Blaabjerg, F., Ionel, D.: Renewable Energy Devices and Systems with Simulation in MATLAB and ANSYS. CRC Press (2017)
12. Clark, C., DuPont, B.: Reliability-based design optimization in offshore renewable energy systems. Renew. Sustain. Energy Rev. **97**, 390–400 (2018)
13. Baranwal, K., Prakash, P., Yadav, V.K.: A modified bypass circuit for improved reliability of PV module validated with real-time data. IEEE Trans. Device Mater. Reliabil (2023). https://doi.org/10.1109/TDMR.2023.3247809
14. Khatib, T., Elmenreich, W.: Modeling of Photovoltaic System Using MATLAB: Simplified Green Codes. Wiley (2016)
15. Xiao, W.: Photovoltaic Power System: Modeling, Design, and Control. Wiley (2017)
16. Duong, P.L.T., Yang, Q., Park, H., Raghavan, N.:Reliability analysis and design of a single diode solar cell model using polynomial chaos and active subspace. Microelectron. Reliabil. **113477**, 100–101 (2019)

17. Ezzati, G.: A reliability-based design optimization model for electricity power networks. In: Dynamics of Continuous, Discrete and Impulsive Systems. Series B: Applications and Algorithms, vol. 22, pp. 339–357 (2015)
18. Ezzati, G.: On probabilistic constraints in optimization models of electricity power networks. Far East J. Appl. Math. **94**(3), 231–246 (2016)
19. Shang, Y.: Constrained consensus in state-dependent directed multiagent networks. IEEE Trans. Netw. Sci. Engin. **9**(6), 4416–4425 (2022)
20. Artin, S., Salimzadeh, S.: A conjugate gradient direction-based method to evaluate reliability analysis problems. IAENG Int. J. Appl. Math. **52**(3), 659–666 (2022)
21. Ezzati, G., Mammadov, M., Kulkarni, S.: A new reliability analysis method based on the conjugate gradient direction. Struct. Multidiscip. Optim. **51**, 89–98 (2015)

Permutation Tests for the Partial Coefficient in a Multiple Linear Regression

Shunpu Zhang

Abstract Existing permutation tests for the partial coefficient in a multiple linear regression model are mostly stepwise algorithms which are difficult to be analyzed and compared theoretically. In this paper, we propose a unified framework for testing the partial coefficient in a multiple linear regression model. The unified framework provides a theoretical basis for constructing permutation tests for the partial coefficient as well as theoretical justifications why they work. As a result, several permutation tests are proposed based on the unified framework. The unified framework also enables us to compare the performances of the permutation tests theoretically. By showing that the existing permutation tests for the partial coefficient of a multiple linear regression model are special cases of the unified framework, we are able to identify redundant steps and provide simpler algorithms for the existing permutation tests. Recommendations on which methods to use under different situations are provided based on the theoretical and numerical comparisons we have conducted.

Keywords Multiple regression model · Partial coefficient · Permutation test · Type I error rate · Power

1 Introduction

Permutation tests have widespread applications in many fields such as econometrics, psychology, and biology, etc. In this paper, our focus will be on testing the partial coefficient in a multiple regression model. Assume that we have the following multiple linear regression model

$$\boldsymbol{Y} = \boldsymbol{X}_1\boldsymbol{\beta}_1 + \boldsymbol{x}_p\boldsymbol{\beta}_1 + \boldsymbol{\varepsilon}, \tag{1}$$

S. Zhang (✉)
Department of Statistics and Data Science, University of Central Florida, Orlando, FL 32816-2370, USA
e-mail: Shunpu.Zhang@ucf.edu

W. Wang et al. (eds.), *Computational Statistics and Data Intelligence*, Springer Proceedings in Mathematics & Statistics 463,
https://doi.org/10.1007/978-981-97-4438-1_3

where

$$\boldsymbol{Y}=\begin{pmatrix} y_1 \\ \vdots \\ y_n \end{pmatrix}, \quad \boldsymbol{X}_1=\begin{pmatrix} 1 & x_{11} & \cdots & x_{1,p-1} \\ 1 & x_{21} & \cdots & x_{2,p-1} \\ \vdots & \vdots & \vdots & \vdots \\ 1 & x_{n1} & \cdots & x_{n,p-1} \end{pmatrix}, \quad \boldsymbol{\beta}_1=\begin{pmatrix} \beta_0 \\ \beta_1 \\ \vdots \\ \beta_{p-1} \end{pmatrix},$$

$$\boldsymbol{x}_p=\begin{pmatrix} x_{1p} \\ \vdots \\ x_{n,p} \end{pmatrix}, \quad \boldsymbol{\varepsilon}=\begin{pmatrix} \varepsilon_1 \\ \vdots \\ \varepsilon_n \end{pmatrix}$$

and $\varepsilon_1, \cdots \varepsilon_n$ are i.i.d. random variables with mean zero and variance σ^2. The hypothesis of interest is

$$H_0 : \beta_p = 0 \quad \text{vs} \quad H_a : \beta_p \neq 0$$

Testing the partial coefficient in a multiple linear regression model presents a unique challenge due to the involvement of the nuisance parameters. The existing methods for testing the partial coefficient of a multiple linear regression model can be classified into three categories: (1) the permuting the$\boldsymbol{Y}$'s method [1] (the Manly method), (2) the permuting the residuals method, and (3) the permuting the $\boldsymbol{x}_p$'s method. The permuting the residuals method can further be classified into two categories: (1) the permuting the residuals under the reduced model method (i.e., the model under the null hypothesis) [2, 3] and the permuting the residuals under the full model method [4]. Numerical comparisons carried out in [5] reported that (1) the Manly method, the method of Freedman and Lane (the FL method [2]), and the method of ter Braak (the tB method [4]) are asymptotically equivalent and have significantly higher power than the normal theory t-test when the errors are non-normal; (2) the Kennedy method [3] has serious inflated type I error problem, and does not perform as well as the FL method; (3) the Manly method is not robust to extreme outliers in the covariates while the FL method and the tB method are more resistant to outliers in the covariates and provide similar results. For the permuting$\boldsymbol{x}_p$'s method, it has been shown that it does not work properly since shuffling $\boldsymbol{x}_p$ changes the collinearity between $\boldsymbol{X}_1$ and $\boldsymbol{x}_p$ [3, 4, 6]. However, O'Gorman [7] argues that the permuting the $\boldsymbol{x}_p$'s method is still a valuable method for testing the partial coefficient in a multiple linear regression model. A more recent review of the permutation tests for regression model can be found in Giacalone and Alibrandi [8]. Permutation testing in high-dimensional linear model was discussed in Hemerik et al. [9].

Despite the extensive literatures on various permutation tests, there is still no clear answer on how and why one method works better than another. The difficulty in comparing the permutation tests lies in the fact that most of the existing permutation methods are in the form of stepwise algorithms. In order to fully understand the differences among the existing methods, it is necessary to put all these methods on the same basis (the unified framework), which is the goal of Sect. 2 in which several permutation tests based on the unified framework are proposed. In Sect. 3, we discuss the choice of the test statistics. In Sect. 4, we demonstrate that the existing permutation tests for the partial coefficient in a multiple linear regression model are special cases of the unified framework. Theoretical and numerical comparisons of the existing and the newly proposed permutation tests are conducted in Sect. 5.

2 The Unified Framework for Permutation Tests for Testing the Partial Coefficient

For testing hypothesis about β_p, it can be seen that the multiple regression model (1) is equivalent to the following simple linear regression model

$$(\boldsymbol{I} - \boldsymbol{H}_1)Y = (\boldsymbol{I} - \boldsymbol{H}_1)\boldsymbol{x}_p\beta_p + (\boldsymbol{I} - \boldsymbol{H}_1)\boldsymbol{\varepsilon}, \tag{2}$$

where $\mathbf{H}_1 = \mathbf{X}_1'\left(\mathbf{X}_1'\mathbf{X}_1\right)'\mathbf{X}_1$.

Model (2) is the unified framework we will work on in the remaining of this paper. The unified framework can greatly enhance our understanding of the permutation tests and facilitate the comparison of different permutation methods. In the following, we propose several permutation tests based on (2) for testing the partial coefficient of a multiple regression model.

2.1 The Permuting the Y's Method

The permuting the $\boldsymbol{Y}$'s method is simply the method of permuting the observed values in the response variable. It is obvious that the permuting the $\boldsymbol{Y}$'s method based on (2) can be constructed as follows:

$$(\boldsymbol{I} - \boldsymbol{H}_1)[Y]^{perm} = (\boldsymbol{I} - \boldsymbol{H}_1)\boldsymbol{x}_p\beta_p + \boldsymbol{\varepsilon}^*, \tag{3}$$

where $\boldsymbol{\varepsilon}^*$ denotes the random error term. In the remaining of the paper, $\boldsymbol{\varepsilon}^*$ will be used to represent the error term in a permutation regression model, and it may represent different random errors in different models.

We need to point out that the permuting the $\boldsymbol{Y}$'s method (3) is not exact due to the fact that the permuted quantities are first multiplied by $(\boldsymbol{I} - \boldsymbol{H}_1)$ before being regressed on $(\boldsymbol{I}-\boldsymbol{H}_1)\boldsymbol{x}_p$, and hence it does not satisfy the exchangeability assumption anymore [10, 11].

2.2 The First Order Correction Method

The unified framework (2) shows that it is more natural to permute $(\boldsymbol{I} - \boldsymbol{H}_1)\boldsymbol{Y}$ and hence a permutation test for $H_0 : \beta_p = 0$ can be conducted based on

$$[(\boldsymbol{I} - \boldsymbol{H}_1)\boldsymbol{Y}]^{perm} = (\boldsymbol{I} - \boldsymbol{H}_1)\boldsymbol{x}_p\beta_p + \boldsymbol{\varepsilon}^*. \tag{4}$$

Notice further that $(\boldsymbol{I} - \boldsymbol{H}_1)$ is idempotent, multiply $(\boldsymbol{I} - \boldsymbol{H}_1)$ on both sides of (4), we obtain a permutation test based on the following model

$$(\boldsymbol{I} - \boldsymbol{H}_1)[(\boldsymbol{I} - \boldsymbol{H}_1)\boldsymbol{Y}]^{perm} = (\boldsymbol{I} - \boldsymbol{H}_1)\boldsymbol{x}_p\beta_p + \boldsymbol{\varepsilon}^*. \tag{5}$$

It can be seen that (4) and (5) permute the same quantities. But the difference is that in (5) the permuted quantities are further multiplied by $(\boldsymbol{I} - \boldsymbol{H}_1)$ before being regressed against $(\boldsymbol{I} - \boldsymbol{H}_1)\boldsymbol{x}_p$. Since $E[(\boldsymbol{I} - \boldsymbol{H}_1)\boldsymbol{Y}] = 0$ under H_0, i.e., the proposed methods correct the first order moment of the permuted quantities to zero, we call (4) and (5) the first order correction method and the modified first order correction method, respectively. The permutation tests (4) and (5) are also not exact since the quantities $(\boldsymbol{I} - \boldsymbol{H}_1)\boldsymbol{Y}$ being permuted do not satisfy the exchangeability assumption.

2.3 The Second Order Correction Method

Note that the non-exact problem of the first order correction methods is caused by the fact that the permuted quantities $(I - \boldsymbol{H}_1)\boldsymbol{Y}$ are not i.i.d. despite that they have the same mean $E[(\boldsymbol{I} - \boldsymbol{H}_1)\boldsymbol{Y}] = 0$ under H_0. The idempotent property of $\boldsymbol{I} - \boldsymbol{H}_1$ and the fact that $Var((\boldsymbol{I} - \boldsymbol{H}_1)\boldsymbol{\varepsilon}) = \sigma^2(\boldsymbol{I} - \boldsymbol{H}_1)$ show that we can use the spectral decomposition of $\boldsymbol{I} - \boldsymbol{H}_1$ to transform the permuted quantities to have mean zero, to be un-correlated and have the same variance. Denote by $\boldsymbol{M}_1$ the spectral decomposition of $\boldsymbol{I} - \boldsymbol{H}_1$, we have

$$\boldsymbol{I} - \boldsymbol{H}_1 = \boldsymbol{M}_1\boldsymbol{M}_1' \quad \text{and} \quad \boldsymbol{M}_1'\boldsymbol{M}_1 = I_{n'},$$

where $n' = n - p$. By multiplying $\boldsymbol{M}_1'$ to both sides of (2), we have

$$\boldsymbol{M}_1'Y = \boldsymbol{M}_1'x_p\beta_p + \boldsymbol{M}_1'\varepsilon. \tag{6}$$

It can be proved that $E(\boldsymbol{M}_1'\boldsymbol{Y}) = 0$ under H_0 and $Var(\boldsymbol{M}_1'\boldsymbol{Y}) = \sigma^2\boldsymbol{I}_{n'}$. Hence, Eq. (6) ensures that that $\boldsymbol{M}_1'\boldsymbol{Y}$ are from a population with the same mean, the same variance, and are uncorrelated. A permutation test for $H_0 : \beta_p = 0$ can then be carried out by regressing the permuted $\boldsymbol{M}_1'\boldsymbol{Y}$ on $\boldsymbol{M}_1'\boldsymbol{x}_p$, i.e.,

$$[\boldsymbol{M}_1'\boldsymbol{Y}]^{perm} = \boldsymbol{M}_1'\boldsymbol{x}_p\beta_p + \boldsymbol{\varepsilon}^* \tag{7}$$

Multiplying both sides of (7) by $\boldsymbol{M}_1$, Eq. (7) is equivalent to

$$\boldsymbol{M}_1[\boldsymbol{M}_1'\boldsymbol{Y}]^{perm} = (\boldsymbol{I} - \boldsymbol{H}_1)\boldsymbol{x}_p\beta_p + \boldsymbol{\varepsilon}^*. \tag{8}$$

Since it corrects the problem up to the second order moment, we call the permutation test based on (8) the second order correction method. It can be seen that this method is exact if the data are from normal populations since having the same mean and the variance in addition to un-correlatedness implies i.i.d. under the normal assumption and hence satisfying the exchangeability condition.

2.4 The First Order Correction Method Based on the Full Model

In the first order correction methods (4) and (5), we see that the quantities being permuted $(\boldsymbol{I} - \boldsymbol{H}_1)\boldsymbol{Y}$ are the residuals from fitting the reduced model $\boldsymbol{Y} = \boldsymbol{X}_1\boldsymbol{\beta}_1 + \boldsymbol{\varepsilon}$. In other words,$(\boldsymbol{I} - \boldsymbol{H}_1)\boldsymbol{Y} = \boldsymbol{Y} - \boldsymbol{X}_1\widehat{\boldsymbol{\beta}}_1$, where $\widehat{\boldsymbol{\beta}}_1$ is the OLS estimator from $\boldsymbol{Y} = \boldsymbol{X}_1\boldsymbol{\beta}_1 + \boldsymbol{\varepsilon}$. This reminds us that we can also permute the residuals $\boldsymbol{Y} - \boldsymbol{X}_1\widehat{\boldsymbol{\beta}}_1$ for $\widehat{\boldsymbol{\beta}}_1$ obtained from regressing $\boldsymbol{Y}$ jointly on $\boldsymbol{X}_1, \boldsymbol{x}_p$. Denote $\boldsymbol{H}_2 = \boldsymbol{x}_p(\boldsymbol{x}_p'\boldsymbol{x}_p)^{-1}\boldsymbol{x}_p'$ and $\boldsymbol{H}_{12} = \boldsymbol{X}_1(\boldsymbol{x}_1'(\boldsymbol{I} - \boldsymbol{H}_2)\boldsymbol{X}_1)^{-1}\boldsymbol{x}_1'(\boldsymbol{I} - \boldsymbol{H}_2)$, and further note the fact that $(\boldsymbol{I} - \boldsymbol{H}_{12})\boldsymbol{x}_p = \boldsymbol{x}_p$, we have

$$(\boldsymbol{I} - \boldsymbol{H}_{12})\boldsymbol{Y} = \boldsymbol{x}_p\beta_p + (\boldsymbol{I} - \boldsymbol{H}_{12})\boldsymbol{\varepsilon}.$$

A permutation test for $H_0 : \beta_p = 0$ can be constructed from the following model

$$[(\boldsymbol{I} - \boldsymbol{H}_{12})\boldsymbol{Y}]^{perm} = \boldsymbol{x}_p\beta_p + \boldsymbol{\varepsilon}^*. \tag{9}$$

Multiplying $\boldsymbol{I} - \boldsymbol{H}_1$ on both sides of (9), the permutation test can be conducted based on

$$(\boldsymbol{I} - \boldsymbol{H}_1)[(\boldsymbol{I} - \boldsymbol{H}_{12})\boldsymbol{Y}]^{perm} = (\boldsymbol{I} - \boldsymbol{H}_1)\boldsymbol{x}_p\beta_p + \boldsymbol{\varepsilon}^*, \tag{10}$$

where we used the property $(\boldsymbol{I} - \boldsymbol{H}_1)(\boldsymbol{I} - \boldsymbol{H}_{12}) = \boldsymbol{I} - \boldsymbol{H}_1$.

2.5 The Null Statistic-Based Permutation Test

In the permutation methods we discussed previously, the permuted quantities are $\boldsymbol{Y} - \boldsymbol{X}_1\widehat{\boldsymbol{\beta}}_1$, where $\widehat{\boldsymbol{\beta}}_1$ is the OLS estimator of $\boldsymbol{\beta}_1$ obtained either from regressing $\boldsymbol{Y}$ on $\boldsymbol{X}_1$, or from regressing $\boldsymbol{Y}$ jointly on $\boldsymbol{X}_1, \boldsymbol{x}_p$. Naturally we want to see if a permutation test can be constructed by permuting the residuals $\widehat{\boldsymbol{\varepsilon}} = \boldsymbol{Y} - \boldsymbol{X}_1\widehat{\boldsymbol{\beta}}_1 - \boldsymbol{x}_p\widehat{\beta}_p$ under the full model.

Denote $H = X(X'X)^{(-1)}X'$, $\hat{\varepsilon}$ can be written as

$$\widehat{\boldsymbol{\varepsilon}} = (\boldsymbol{I} - \boldsymbol{H})\boldsymbol{Y}.$$

Multiplying $\boldsymbol{I} - \boldsymbol{H}$ to both sides of (1) from the left

$$(\boldsymbol{I} - \boldsymbol{H})\boldsymbol{Y} = (I - \boldsymbol{H})\boldsymbol{\varepsilon},$$

We see that a permutation test for H_0 cannot directly be constructed by permuting $(\boldsymbol{I} - \boldsymbol{H})\boldsymbol{Y}$ since it is independent of β_p. However, note that the OLS estimator $\widehat{\beta}_p$ of β_p can written as

$$\begin{aligned}\hat{\beta}_p &= \left[\boldsymbol{x}_p'(\boldsymbol{I} - \boldsymbol{H}_1)\boldsymbol{x}_p\right]^{-1}\boldsymbol{x}_p'(\boldsymbol{I} - \boldsymbol{H}_1)\boldsymbol{Y} \\ &= \beta_p + \left[\boldsymbol{x}_p'(\boldsymbol{I} - \boldsymbol{H}_1)\boldsymbol{x}_p\right]^{-1}\boldsymbol{x}_p'(\boldsymbol{I} - \boldsymbol{H}_1)\boldsymbol{\varepsilon}.\end{aligned}$$

Hence, $[\boldsymbol{x}_p'(\boldsymbol{I} - \boldsymbol{H}_1)\boldsymbol{x}_p]^{-1}\boldsymbol{x}_p'(\boldsymbol{I} - \boldsymbol{H}_1)\boldsymbol{\varepsilon}$ has the distribution of $\widehat{\beta}_p$ under H_0 (i.e., the null distribution). Since $(\boldsymbol{I} - \boldsymbol{H})\boldsymbol{Y}$ is the residual vector from regressing $\boldsymbol{Y}$ jointly on $\boldsymbol{X}_1, \boldsymbol{x}_p$, it can therefore be used as an estimate of $\boldsymbol{\varepsilon}$. Hence, $[\boldsymbol{x}_p'(\boldsymbol{I} - \boldsymbol{H}_1)\boldsymbol{x}_p]^{-1}\boldsymbol{x}_p'(\boldsymbol{I} - \boldsymbol{H}_1)(\boldsymbol{I} - \boldsymbol{H})\boldsymbol{Y}$ can be used as a null statistic and its distribution can be used as an approximation to the null distribution of $\widehat{\beta}_p$. Furthermore, note that $[\boldsymbol{x}_p'(\boldsymbol{I} - \boldsymbol{H}_1)\boldsymbol{x}_p]^{-1}\boldsymbol{x}_p'(\boldsymbol{I} - \boldsymbol{H}_1)(\boldsymbol{I} - \boldsymbol{H})\boldsymbol{Y}$ is the OLS estimator of β_p from regressing $(\boldsymbol{I} - \boldsymbol{H})\boldsymbol{Y}$ jointly on $\boldsymbol{X}_1, \boldsymbol{x}_p$, a permutation test for H_0 can be constructed by comparing the value of $\widehat{\beta}_p$ with the OLS estimates $\widehat{\beta}_p^*$ obtained from

$$[(\boldsymbol{I} - \boldsymbol{H})\boldsymbol{Y}]^{perm} = \boldsymbol{X}_1\boldsymbol{\beta}_1 + \boldsymbol{x}_p\beta_p + \boldsymbol{\varepsilon}^*,$$

or equivalently from

$$(\boldsymbol{I} - \boldsymbol{H}_1)[(I - \boldsymbol{H})\boldsymbol{Y}]^{perm} = (\boldsymbol{I} - \boldsymbol{H}_1)\boldsymbol{x}_p\beta_p + \boldsymbol{\varepsilon}^*. \tag{11}$$

Since the above permutation test is based on the null statistic. We call it the null statistic-based permutation test. Note that the OLS estimate $\widehat{\beta}_p^*$ from (11) for the un-permuted data is zero, instead of $\widehat{\beta}_p$, the OLS estimator from the un-permuted data. The permutation test defined by (11) is not a conventional permutation test.

3 The Choice of the Test Statistics

In this section we discuss the choice of test statistics. For convenience, we write the test statistics (the OLS estimators of β_p) from the previously discussed permutation models as follows.

The permuting the $\boldsymbol{Y}$'s method

$$\hat{\beta}_p^0 = \left[\boldsymbol{x}_p'(\boldsymbol{I} - \boldsymbol{H}_1)\boldsymbol{x}_p\right]^{-1}\boldsymbol{x}_p'(\boldsymbol{I} - \boldsymbol{H}_1)[(\boldsymbol{I} - \boldsymbol{H}_1)\boldsymbol{Y}]^{perm}, \tag{12}$$

The first order correction method

$$\hat{\beta}_p^1 = \left[\boldsymbol{x}_p'(\boldsymbol{I} - \boldsymbol{H}_1)\boldsymbol{x}_p\right]^{-1}\boldsymbol{x}_p'(\boldsymbol{I} - \boldsymbol{H}_1)[(\boldsymbol{I} - \boldsymbol{H}_1)\boldsymbol{Y}]^{perm}, \tag{13}$$

The modified first order correction method

$$\hat{\beta}_p^{m1} = \left[\boldsymbol{x}_p'(\boldsymbol{I} - \boldsymbol{H}_1)\boldsymbol{x}_p\right]^{-1}\boldsymbol{x}_p'(\boldsymbol{I} - \boldsymbol{H}_1)[(\boldsymbol{I} - \boldsymbol{H}_1)\boldsymbol{Y}]^{perm} \tag{14}$$

The second order correction method

$$\hat{\beta}_p^2 = \left[\boldsymbol{x}_p'(\boldsymbol{I} - \boldsymbol{H}_1)\boldsymbol{x}_p\right]^{-1}\boldsymbol{x}_p'(\boldsymbol{I} - \boldsymbol{H}_1)\boldsymbol{M}_1\left[\boldsymbol{M}_1'\boldsymbol{Y}\right]^{perm} \tag{15}$$

The first order correction method based on the full model

$$\hat{\beta}_p^{1F} = \left[\boldsymbol{x}_p'(\boldsymbol{I} - \boldsymbol{H}_1)\boldsymbol{x}_p\right]^{-1}\boldsymbol{x}_p'(\boldsymbol{I} - \boldsymbol{H}_1)[(\boldsymbol{I} - \boldsymbol{H}_{12})\boldsymbol{Y}]^{perm} \tag{16}$$

The null-statistic based permutation test:

$$\hat{\beta}_p^{H_0} = \left[\boldsymbol{x}_p'(\boldsymbol{I} - \boldsymbol{H}_1)\boldsymbol{x}_p\right]^{-1}\boldsymbol{x}_p'(\boldsymbol{I} - \boldsymbol{H}_1)[(\boldsymbol{I} - \boldsymbol{H})\boldsymbol{Y}]^{perm} \tag{17}$$

It can be seen from (12)–(16) that, for the un-permuted data, each test statistic above is the same as $\widehat{\beta}_p$, the original OLS estimator of β_p from (1). For $\widehat{\beta}_p^{H_0}$ defined by (17), its value from the un-permuted data is zero.

It has been discussed in Sect. 2 that the permuted quantities (12)–(17) do not satisfy the exchangeability assumption. Therefore, all the resulting permutation tests are not exact. In such a case, a t-type test statistic becomes necessary unless the t is a monotonic function of $\widehat{\beta}$, in which case t and $\widehat{\beta}$ are called "equivalent statistics for permutation tests" and the t-type test statistic is called a pivotal statistic, see [12, 13]. Numerous researches have confirmed that the permutation tests based on the pivotal statistic have better control of type I error rate than that based on the non-pivotal test statistic [14, 15].

Based on the unified framework, the t-type test statistic for testing hypothesis $H_0 : \beta_p = 0$ from the un-permuted data is

$$t = \frac{\widehat{\beta}}{se(\widehat{\beta})}, \tag{18}$$

and the t-type test statistic based on the permuted data is

$$t^* = \frac{\widehat{\beta}^*}{se(\widehat{\beta}^*)}, \tag{19}$$

where $se(\hat{\beta}) = \sqrt{\frac{SSE}{df} / \left[x_p'(\boldsymbol{I} - \boldsymbol{H}_1)x_p\right]}$ and $se(\hat{\beta}^*) = \sqrt{\frac{SSE^*}{df} / \left[x_p'(\boldsymbol{I} - \boldsymbol{H}_1)x_p\right]}$, and SSE, SSE^* are the sum of squared errors obtained from the un-permuted and permuted regression models with the unified regressor $(\boldsymbol{I} - \boldsymbol{H}_1)\boldsymbol{x}_p$, and df is the corresponding degrees of freedom.

In the following we derive the *SSE*s and their corresponding degrees of freedom to be used in (19) for each permutation method based on the unified framework.

Theorem 1 *Let* $\boldsymbol{Q} = I - (\boldsymbol{I} - \boldsymbol{H}_1)\boldsymbol{x}_p[\boldsymbol{x}_p'(\boldsymbol{I} - \boldsymbol{H}_1)\boldsymbol{x}_p]^{-1}\boldsymbol{x}_p'(\boldsymbol{I} - \boldsymbol{H}_1)$.

The *SSE*s corresponding to (13)–(18) are as follows:

1. For the permuting $\boldsymbol{Y}$'s method (3),

$$SSE^0 = \hat{\boldsymbol{\varepsilon}}_0'\boldsymbol{Q}'(\boldsymbol{I} - \boldsymbol{H}_1)\boldsymbol{Q}\hat{\varepsilon}_0, \tag{20}$$

   where $\hat{\boldsymbol{\varepsilon}}_0 = (\boldsymbol{I} - \boldsymbol{H}_1)[\boldsymbol{Y}]^{perm}$.
2. For the first order correction method and the modified first order correction methods (4) and (5),

$$SSE^1 = \{[\hat{\boldsymbol{\varepsilon}}_1]^{perm}\}'\boldsymbol{Q}'Q[\hat{\boldsymbol{\varepsilon}}_1]^{perm}, \tag{21}$$

$$SSE^{m1} = \{[\hat{\boldsymbol{\varepsilon}}_1]^{perm}\}'\boldsymbol{Q}'(\boldsymbol{I} - \boldsymbol{H}_1)\boldsymbol{Q}[\hat{\boldsymbol{\varepsilon}}_1]^{perm}, \tag{22}$$

   where $\hat{\boldsymbol{\varepsilon}}_1 = (\boldsymbol{I} - \boldsymbol{H}_1)\boldsymbol{Y}$.
3. For the second order correction method (8),

$$SSE^2 = \hat{\boldsymbol{\varepsilon}}_2'\boldsymbol{Q}'(\boldsymbol{I} - \boldsymbol{H}_1)\boldsymbol{Q}\hat{\varepsilon}_2, \tag{23}$$

   where $\hat{\boldsymbol{\varepsilon}}_2 = \boldsymbol{M}_1[\boldsymbol{M}_1'(\boldsymbol{I} - \boldsymbol{H}_1)\boldsymbol{Y}]^{perm}$.
4. For the first order correction method based on the full model (11),

$$SSE^{1F} = \{[\hat{\boldsymbol{\varepsilon}}_{12}]^{perm}\}'\boldsymbol{Q}'(\boldsymbol{I} - \boldsymbol{H}_1)\boldsymbol{Q}[\hat{\boldsymbol{\varepsilon}}_{12}]^{perm}, \tag{24}$$

where $\widehat{\boldsymbol{\varepsilon}}_{12} = (\boldsymbol{I} - \boldsymbol{H}_{12})\boldsymbol{Y}$.

5. For the null-statistic based permutation test (10),

$$SSE^{H_0} = \{[\widehat{\boldsymbol{\varepsilon}}]^{perm}\}'\boldsymbol{Q}'(\boldsymbol{I} - \boldsymbol{H}_1)\boldsymbol{Q}[\widehat{\boldsymbol{\varepsilon}}]^{perm}, \tag{25}$$

where $\widehat{\boldsymbol{\varepsilon}} = (\boldsymbol{I} - \boldsymbol{H})\boldsymbol{Y}$.

Further, we have the following results:

(a) All the above *SSE*s have degrees of freedom $n-p-1$ except for SSE^1, the *SSE* of the first order correction method, whose degrees of freedom is $n-p$.
(b) For the un-permuted data, the above SSEs are the same and are equal to $Y'(\boldsymbol{I} - \boldsymbol{H})Y$, the *SSE* from fitting the multiple regression model (1).

The proof of Theorem 1 is deferred to the full paper, which can be requested from the author.

Remark 1 Theorem 1 also reveals that, for the first order correction method (4), the degrees of freedom of the SSE with respect to the un-permutated regression model is $n-p-1$, but that of the SSE from the permuted regression model is $n-p$.

Remark 2 Among all the permutation methods we discussed above, only for two methods (the first order correction method and the second order correction method), t^* is a monotonic function of $\widehat{\beta}^*$. However, since the first order correction method uses different degrees of freedom in the test statistics t and t^*, the only method which has "equivalent test statistic" is the second order correction method. In other words, the second order correction method does not require using the pivotal statistic (i.e., the t-statistic) in testing.

The proposed permutation tests based on the t-type test statistic (18) and (19) can be conducted as follows:

Step 1. Obtain the value of the t-statistic $t = \widehat{\beta}_p/s(\widehat{\beta}_p)$ for testing $H_0 : \beta_p = 0$ from (2), where $\widehat{\beta}_p$ is the OLS estimate of β_p and $s(\widehat{\beta}_p)$ is its standard error with df $= n-p-1$.
Step 2: Permute the quantities specified in $[\cdot]^{perm}$ on the left hand side of the permutation model (3)–(5), (8), (10), or (11). Then, obtain the OLS estimate $\widehat{\beta}_p^*$ and the standard error $se(\widehat{\beta}_p^*)$ from the corresponding permutation model. The degrees of freedom used in $se(\widehat{\beta}_p^*)$ are $n-p-1$ for all proposed methods except for the first order correction method (4) whose df is $n-p$.

The permutation distribution is obtained by calculating $t^* = \widehat{\beta}_p^*/se(\widehat{\beta}_p^*)$ for all the permutations and H_0 is rejected if t is beyond the $\alpha/2$ or $1-\alpha/2$ percentiles of the permutation distribution.

The above algorithm requires matrix manipulation to obtain the permuted quantities. In practice, it is often desired that the matrix manipulation can be avoided if possible. Note that, for the purpose of estimating β_p, the model $(\boldsymbol{I} - \boldsymbol{H}_1)[\cdot]^{perm} =$

$(\boldsymbol{I} - \boldsymbol{H}_1)\boldsymbol{x}_p\beta_p + \boldsymbol{\varepsilon}^*$ is equivalent to regressing $[\cdot]^{perm}$ on $\boldsymbol{X}_1\boldsymbol{\beta}_1 + \boldsymbol{x}_p\beta_p$, the regressor of the full model (1). In the following we provide alternative algorithms for the permutation tests we proposed.

3.1 The Permuting the Y'S Method

Step 1. Obtain the value of the t-statistic $t = \widehat{\beta}_p/s(\widehat{\beta}_p)$ for testing $H_0 : \beta_p = 0$ from (1), where $\widehat{\beta}_p$ is the OLS estimate of β_p and $s(\widehat{\beta}_p)$ is its standard error.
Step 2: Permute $\boldsymbol{Y}$, and denote it by $[\boldsymbol{Y}]^{perm}$. Obtain the OLS estimate $\widehat{\beta}_p^*$ and the standard error $se(\widehat{\beta}_p^*)$ from

$$[\boldsymbol{Y}]^{perm} = \boldsymbol{X}_1\boldsymbol{\beta}_1 + \boldsymbol{x}_p\beta_p + \boldsymbol{\varepsilon}^*.$$

The permutation distribution is obtained by calculating $t^* = \widehat{\beta}_p^*/se(\widehat{\beta}_p^*)$ for many permutations. The degrees of freedom used in the calculation of t and t^* are $n-p-1$.

Using the fact that $(\boldsymbol{I} - \boldsymbol{H}_1)\boldsymbol{Y}$ is the residual vector from regressing $\boldsymbol{Y}$ on $\boldsymbol{Y} = \boldsymbol{\beta}_0 + \boldsymbol{X}_1\boldsymbol{\beta}_1 + \boldsymbol{\varepsilon}$, we have the following stepwise algorithm for the first order correction method.

3.2 The First Order Correction Method

Step 1: The same as the permuting the $\boldsymbol{Y}$'s method.
Step 2: Obtain the residual vector from regressing $\boldsymbol{Y}$ on $\boldsymbol{Y} = \boldsymbol{\beta}_0 + \boldsymbol{X}_1\boldsymbol{\beta}_1 + \boldsymbol{\varepsilon}$, and denote it by $\widehat{\boldsymbol{\varepsilon}}_1$.
Step 3: Permute $\widehat{\boldsymbol{\varepsilon}}_1$ and obtain the OLS estimate $\widehat{\beta}_p^*$ and the standard error $se(\widehat{\beta}_p^*)$ from regressing the permuted $\widehat{\boldsymbol{\varepsilon}}_1$ on $(I - \boldsymbol{H}_1)\boldsymbol{x}_p$, i.e.,

$$[\widehat{\boldsymbol{\varepsilon}}_1]^{perm} = (I - \boldsymbol{H}_1)\boldsymbol{x}_p\beta_p + \boldsymbol{\varepsilon}^*.$$

The degrees of freedom used in the calculation of t and t^* are $n-p-1$ and $n-p$, respectively.

For all the other proposed methods, the degrees of freedom used in the calculation of t, t^* are both $n-p-1$.

3.3 The Modified First Order Correction Method

Step 1–2: The same as the first order correction method.
Step 3: Permute $\widehat{\boldsymbol{\varepsilon}}_1$ and obtain the OLS estimate $\widehat{\beta}_p^*$ and the standard error $se(\widehat{\beta}_p^*)$ from regressing $\widehat{\boldsymbol{\varepsilon}}_1$ on the full model (1), i.e.,

$$[\widehat{\boldsymbol{\varepsilon}}_1]^{perm} = \boldsymbol{X}_1\boldsymbol{\beta}_1 + \boldsymbol{x}_p\beta_p + \boldsymbol{\varepsilon}^*.$$

3.4 The Second Order Correction Method

Step 1: The same as the permuting the y's method.
Step 2: Obtain $\boldsymbol{M}_1$ from the spectral decomposition of $\boldsymbol{I} - \boldsymbol{H}_1$ and calculate $\boldsymbol{M}_1'\boldsymbol{Y}$. Denote $\boldsymbol{M}_1'\boldsymbol{Y}$ by $\widehat{\boldsymbol{\varepsilon}}_2$.
Step 3: Permute $\widehat{\boldsymbol{\varepsilon}}_2$ and obtain the OLS estimate $\widehat{\beta}_p^*$ and the standard error $se(\widehat{\beta}_p^*)$ from regressing the permuted $\widehat{\boldsymbol{\varepsilon}}_2$ on the full model (1), i.e.,

$$[\widehat{\boldsymbol{\varepsilon}}_2]^{perm} = \boldsymbol{X}_1\boldsymbol{\beta}_1 + \boldsymbol{x}_p\beta_p + \boldsymbol{\varepsilon}^*.$$

3.5 The First Order Correction Method Based on the Full Model

Step 1: Obtain the OLS estimates $\widehat{\boldsymbol{\beta}}_1$ and $\widehat{\beta}_p$ from (1) and calculate $t = \widehat{\beta}_p/s(\widehat{\beta}_p)$.
Step 2: Obtain the residual vector $\boldsymbol{Y} - \boldsymbol{X}_1\widehat{\boldsymbol{\beta}}_1$ for $\widehat{\boldsymbol{\beta}}_1$ obtained in Step 1 and denote it by $\widehat{\boldsymbol{\varepsilon}}_{12}$.
Step 3: Permute $\widehat{\boldsymbol{\varepsilon}}_{12}$ and obtain the OLS estimate $\widehat{\beta}_p^*$ and the standard error $se(\widehat{\beta}_p^*)$ from regressing the permuted $\widehat{\boldsymbol{\varepsilon}}_{12}$ on the full model (1), i.e.,

$$[\widehat{\boldsymbol{\varepsilon}}_{12}]^{perm} = \boldsymbol{X}_1\boldsymbol{\beta}_1 + \boldsymbol{x}_p\beta_p + \boldsymbol{\varepsilon}^*.$$

3.6 The Null Statistic-Based Permutation Test

Step 1: Fit the full model (1), obtain the residual vector $\widehat{\boldsymbol{\varepsilon}}$ and calculate $t = \widehat{\beta}_p/s(\widehat{\beta}_p)$.
Step 2: Permute $\widehat{\boldsymbol{\varepsilon}}$ and obtain the OLS estimate $\widehat{\beta}_p^*$ and the standard error $se(\widehat{\beta}_p^*)$ from regressing the permuted $\widehat{\boldsymbol{\varepsilon}}$ on the full model (1), i.e.,

$$[\hat{\boldsymbol{\varepsilon}}]^{perm} = \boldsymbol{X}_1\boldsymbol{\beta}_1 + \boldsymbol{x}_p\beta_p + \boldsymbol{\varepsilon}^*.$$

4 The Existing Methods and Their Relationships with the Proposed Methods

The step-wise algorithms of the well-known existing permutation tests for testing partial coefficients in multiple linear regression models are provided in Anderson and Legendre [5]. For convenience of comparison with our proposed methods based on the unified framework, we list these methods here.

1. The Manly method/the permuting the $\boldsymbol{Y}$'s method
2. The Freedman and Lane (FL) method
3. The Kennedy method
4. The ter Braak (tB) method
5. The Huh and Jhun (HJ) method [16]

Comparing the algorithms of the above methods with our proposed methods, we can see that: 1. the Manly method is equivalent to the permuting the $\boldsymbol{Y}$'s method (3); 2. the HJ method is the same as the second order correction method (8). However, for the other methods, the stepwise form of the algorithms makes it difficult to see their relationships with our proposed methods. Through a theoretical treatment to put these methods on the unified framework we proposed in Sect. 2, we have the following findings:

(a) the FL method is equivalent to the modified first order correction method, but the step to add the fitted values $\boldsymbol{H}_1\boldsymbol{Y}$ to the permuted residuals in the FL method is un-necessary;
(b) there is no counterpart of the first order correction method based on the full model in the existing methods;
(c) the Kennedy method uses the same permutation model as the first order correction method. But the degrees of freedom used in t and t^* are different for the two methods: we use $n-p-1$ and $n-p$ as the degrees of freedom of t and t^*, while the Kennedy's method uses the same degree of freedom $n-p-1$ for both.
(d) the null statistics-based permutation test (11) is equivalent to the tB test, but the step to add the fitted values $\boldsymbol{HY}$ to the permuted residuals in the tB method is un-necessary.

Details of the comparison of the above methods with our proposed methods can be found in the full paper.

5 The Performance of the Permutation Tests

5.1 Theoretical Comparison

Ideally, we hope that we can provide an exact ranking on the performance of the permutation methods. However, this is only possible if the OLS estimators have the same permutation distribution, in which case a comparison of the performance of the permutation methods based on the pivotal statistic can be conducted by comparing their respective mean squared error (MSE) values. Among all the methods discussed in this paper, there are only three tests meeting this requirement: the first order correction method (4), the modified first order correction method (5), and the Kennedy method, which share the common OLS estimator.

$$\hat{\beta}_p^1 = \hat{\beta}_p^{m1} = \left[\boldsymbol{x}_p'(\boldsymbol{I} - \boldsymbol{H}_1)\boldsymbol{x}_p\right]^{-1}\boldsymbol{x}_p'(\boldsymbol{I} - \boldsymbol{H}_1)\left[(\boldsymbol{I} - \boldsymbol{H}_1)\boldsymbol{Y}\right]^{perm}.$$

For comparing these three methods, we derive the following result.

Theorem 2 *For SSE^1, SSE^{m1} defined in Theorem 1, we have*

$$SSE^1 \geq SSE^{m1}.$$

The proof of Theorem 2 can be found in the appendix. Theorem 2 shows that

(1) Note that $SSE^1/(n-p)$ is an unbiased estimator of σ^2 for the unpermuted data, the MSE used in the Kennedy method ($SSE^1/(n-p-1)$) over-estimates σ^2. A direct result of this is the inflated type I error rate problem of the Kennedy method, which has been reported in numerous literatures.
(2) Since $SSE^1/(n-p-1) > SSE^1/(n-p)$ (the latter is the MSE used in the first order correction method), the Kennedy method always has higher probability of rejecting H_0 than the first order correction method.
(3) Since $SSE^1/(n-p-1) \geq SSE^{m1}/(n-p-1)$), where $SSE^{m1}/(n-p-1)$ is the MSE used in the modified first order correction method, the Kennedy method always has higher probability of rejecting H_0 than the modified first order correction method.
(4) Note that we have $SSE^1/(n-p) \geq SSE^{m1}/(n-p-1)$ if the sample size n satisfies $1/(n-p-1) \approx 1/(n-p)$. Then, the first order correction method has higher probability of rejecting H_0 than the modified first order correction method.

For the other permutation methods we have discussed in this paper, a comparison of the MSEs will not tell us which method works better due to the fact that the test statistics for different methods have different permutation distributions, which are theoretically un-tractable without knowing the true error distribution. For large sample sizes, we expect that they will have similar performances since the t-type test statistics for all the methods except the Kennedy method have the same degrees of freedom $n-p-1$, see Anderson and Robinson [17]. To evaluate their finite sample performances, we resort to numerical methods to compare these methods in the following sub-section.

5.2 *Numerical Comparison*

There have been extensive simulation results reported in the literatures for the existing permutation methods [5, 18, 19], among others. Due to the inclusion of new permutation methods, we have conducted simulations for all the methods we discussed in this paper using a setup similar to that used [5].

The simulations are organized as follows. We first investigate the performances of the permutation methods with respect to the normal error. Then we will examine the performances of the permutation methods when the covariate $\boldsymbol{X}_1$ contains outliers and when the error distribution deviates from the normal distribution. We use two extreme distributions: the Cauchy and the exponential distributions in our simulations, where the latter means the distribution of an exponential random variable cubed.

It is well known that the testing problem in simple linear regression through the origin is not robust to non-constant variance of the error variable (heteroscedasticity). The equivalence between the problem of testing the partial coefficient in a multiple regression and the problem of testing the slope of a simple linear regression model through the origin shows that we need to examine the performances of the permutation methods for non-constant error variances, a situation which has not been considered [5, 18].

In the simulations, the data were generated from the following model

$$\boldsymbol{Y} = (\boldsymbol{X}_1, \boldsymbol{x}_2)\boldsymbol{\beta} + \boldsymbol{\varepsilon}, \tag{26}$$

where $\boldsymbol{X}_1, \boldsymbol{x}_2$ are set uniformly at values 1, 2, 3 in a crossed 3×3 design, $\boldsymbol{\beta} = (\beta_0, \beta_1, \beta_2)'$ and $\boldsymbol{\varepsilon}$ is the random error vector. Throughout the simulations, we choose $\beta_0 = \beta_1 = 1$. The nominal type I error rate is chosen as $\alpha = 0.05$.

We have also conducted simulation to investigate

1. The effect of outliers in the covariate $\boldsymbol{X}_1$ on the control of the type I error rate
2. The effect of heteroscedasticity on the control of the type I error rate
3. The robustness with respect to the deviations from the normal error
4. The effect of correlation on the control of the type I error rate

Detailed simulation results can be found in the full paper.

5.3 Conclusions and Discussions

The numerical comparisons we have carried out show that (1) the t-test has unreliable control of the type I error rate when the error has non-constant variances or has extremely skewed distributions; (2) the Kennedy method has the inflated type I error rate problem; (3) the use of the permuting the $\boldsymbol{Y}$'s method should be avoided when there are outliers in the covariate; (4) the HJ method has good control of the type I error rate when the error are not normal, but its power is lower than the other methods (despite that they have the deflated control of the type I error rate) and is extremely sensitive to non-constant error variances; (5) the null statistics-based method is not robust to both non-constant error variances and non-normal errors; (6) the first order correction method is robust to non-constant variances, but its control of the type I error rate may be slightly affected by the correlation between $\boldsymbol{X}_1, \boldsymbol{x}_2$; (7) the modified first order correction method and the first order correction method based on the full model have almost equivalent performances, however the first order correction method based on the full model is not robust to correlated $\boldsymbol{X}_1, \boldsymbol{x}_2$. Overall, we recommend using the first order correction method when the normality assumption or the constant variance assumption is violated and using the modified first order correction method in other cases.

References

1. Manly, B.F.J.: Randomization, Bootstrap, and Monte Carlo Methods in Biology, 3rd edn., Chapman & Hall, London (2007)
2. Freedman, D., Lane, D.: A nonstochastic interpretation of reported significance levels. J. Bus. Econ. Stat. **1**, 292–298 (1983)
3. Kennedy, P.E.: Randomization tests in econometrics. J. Bus. Econ. Stat. **13**, 85–94 (1995)
4. ter Braak, C.J.F.: Permutation versus bootstrap significance tests in multiple regression and ANOVA. In: Jockel, K.J. (ed.) Bootstrapping and Related Techniques, pp. 79–86. Springer, Berlin (1992)
5. Anderson, M.J., Legendre, P.: An empirical comparison of permutation methods for tests of partial regression coefficients in a linear model. J. Stat. Comput. Simul. **62**(3), 271–303 (1999)
6. Welch, W.J.: Construction of permutation tests. J. Am. Statist. Ass. **85**, 693–698 (1990)
7. O'Gorman, T.W.: The performance of randomization tests that use permutations of independent variables. Commun. Stat.—Simul. Comput. **34**(4), 895–908 (2007)
8. Giacalone, M., Alibrandi, A.: Overview and main advances in permutation tests for linear regression models. J. Math. Syst. Sci. **5**, 53–59 (2015)
9. Hemerik, J., Thoresenb, M., Finosc, L.: Permutation testing in high-dimensional linear models: an empirical investigation. J. Stat. Comput. Simul. **91**(5), 897–914 (2021)
10. Good, P.: Permutation Tests: A Practical Guide to Resampling Methods for Testing Hypotheses. Springer, New York (1994)
11. Zhang, S.: The split sample permutation t-tests. J. Stat. Plann. Inferences **39**(10), 3512–3524 (2009)
12. Edgington, E.S.: Randomizsation Tests, 3rd edn. Marcel Dekker, New York (1995)
13. Edgington, E.S., Onghena, P.: Randomization Tests, 4th edn. Chapman & Hall, London (2007)
14. Hall, P., Titterington, D.M.: The effect of simulation order on level accuracy and power of Monte Carlo tests. J. R. Statist. Soc. B **51**, 459–467 (1989)

15. Hall, P., Wilson, S.R.: Two guidelines for bootstrap hypothesis testing. Biometrics **47**, 757–762 (1991)
16. Huh, M.H., Jhun, M.: Random permutation testing in multiple regression. Commun. Stat.—Theory Methods **30**(10), 2023–2032 (2001)
17. Anderson, M.J., Robinson, J.: Permutation tests for linear models. Aust. N. Z. J. Stat. **43**(1), 75–88 (2001)
18. Kennedy, P.E., Cade, B.S.: Randomization tests for multiple regression. Comm. Statist. Simul. Comput. **25**, 923–936 (1996)
19. Anderson, M.J.: Permutation tests for univariate or multivariate analysis of variance and regression. Can. J. Fish. Aquat. Sci. **58**, 626–639 (2001)

Research on Population Mortality Based on Generalized Partially Function Linear Model with Interaction Term

Kejing Mao, Songxuan Li, and Weiwei Xiao

Abstract This paper briefly introduces a generalized partially function type linear model with interaction terms and applies the model to the study of population mortality, exploring the effects of air quality index, temperature, GDP per capita and number of beds per 1,000 population in hospitals on mortality in selected Chinese cities in 2019. It was concluded that both the air quality index and temperature were positively correlated with the population mortality rate, while GDP per capita and number of beds per 1,000 population in hospitals were negatively correlated with the population mortality rate.

Keywords Functional data analysis · Interaction items · Mortality rate

1 Introduction

Nowadays, most of the data included are in the form of curves and images, and functional data has a great advantage in dealing with this type of data, so we use functional data analysis to study the relationship between these data. Statistical methods such as regression analysis [1], principal component analysis [2] have been well integrated into functional data analysis, but more research is needed.

Ramsay [3] clarified what functional data are, and discussed the least square approximation and principal component analysis. Ranmsay et al. [4] summarized the basic form of functional linear models and their estimation algorithms. Nelder et al. [5] were the first to introduce the generalized linear model. Muller et al. [6] proposed a linear regression model of generalized function type in view of the regression situation where the response variable is scalar and the predictor variable is a random function. In real data analysis, there is often some relationship between predictor variables. Therefore, many researchers have explored regression models with interaction

K. Mao · S. Li · W. Xiao (✉)
North China University of Technology, Beijing, China
e-mail: xiaoww@ncut.edu.cn

W. Wang et al. (eds.), *Computational Statistics and Data Intelligence*, Springer Proceedings in Mathematics & Statistics 463,
https://doi.org/10.1007/978-981-97-4438-1_4

terms. Yang et al. [7] improved the prediction of spawning rate of sturgeon by considering the functional interaction between depth spectra and temperature time series. Usset [8] proposed a functional regression model with scalar response variables and functional predictor variables that can be adjusted for two-way interaction.

Population mortality is an important statistical indicator of the health status of a population and is strongly linked to genetic, social and environmental factors. Kong et al. [9] examined the effects of PM2.5 concentrations, temperature and other factors on mortality in selected US cities. Ying et al. [10] examined the factors influencing mortality from different perspectives. This paper considers both the role of the link function and the interaction between the function-based predictor variables, based on existing research, which makes the results of practical applications more accurate.

2 Model

Assuming that there is a scalar response variable $Y_i, i = 1, \cdots, n$ and $X_{ij}(t_j), j = 1, 2$ is a functional predictor variable that affects Y_i, where $X_{ij}(t_j) \in L^2(T_j)$, q dimensional vector $Z = (Z_1, Z_2, \cdots, Z_q)$ is a scalar predictor variable that affects Y_i, $g(\cdot)$ is a known link function with continuous second order derivative. Define the response variable Y_i and the predictor variables $X_{ij}(t_j)$ and Z to have the following relationship:

$$
\begin{aligned}
Y_i = g\Bigg(\alpha + \int_{T_1} X_{i1}(t_1)\beta_1(t_1)dt_1 + \int_{T_2} X_{i2}(t_2)\beta_2(t_2)dt_2 \\
+ \iint_{T_1 T_2} X_{i1}(t_1)X_{i2}(t_2)\beta(t_1, t_2)dt_1 dt_2 + Z^T\gamma\Bigg) + \varepsilon_i
\end{aligned}
\tag{1}
$$

where $\alpha \in \mathrm{R}$ is the intercept, $\beta_1(t_1)$, $\beta_2(t_2)$ and $\beta(t_1, t_2)$ are regression coefficient functions, and $\gamma = (\gamma_1, \gamma_2, \cdots, \gamma_q)^T$ is the unknown regression coefficient corresponding to the scalar predictor variable Z. It is assumed that ε_i is independent and has a mean of 0 and a variance of σ^2.

We bring the K-L expansion of $X_{ij}(t_j)$, $\beta_j(t_j)$, $\beta(t_1, t_2)$ into (1) and truncate the predictor variable at p_j to obtain the new model

$$
Y_i = g(\alpha + \sum_{k=1}^{p_1} \theta_{i1k}\omega_{1k} + \sum_{l=1}^{p_2} \theta_{i2l}\omega_{2l} + \sum_{k=1}^{K}\sum_{l=1}^{L} \rho_{ikl}u_{kl} + Z^T\gamma) + \varepsilon_i, \; i = 1, 2, \cdots, n
\tag{2}
$$

where $K = p_1, L = p_2$.

3 Application

This paper uses the proposed model to explore some of the factors that influence mortality. The data used in this paper were obtained from environmental monitoring stations and statistical bulletins of 80 cities including Beijing, Benxi and Cangzhou from 1 January 2019 to 31 December 2019 on air quality index, temperature, GDP per capita, number of beds per 1,000 population in hospitals and mortality.

Among them, the air quality index and temperature are taken as two functional predictor variables, which are denoted as X_1 and X_2 respectively. The GDP per capita and the number of beds per 1,000 population in hospitals are taken as two scalar predictor variables, which are denoted as Z_1 and Z_2 respectively. The mortality rate is taken as a response variable, which is denoted as Y. We first pre-processed the mortality data, and if the mortality rate is greater than 6‰, the city is considered to have a high mortality rate, denoted by 1, and conversely, a low mortality rate, denoted by 0. Figure 1 shows the air quality index and temperature for selected cities in 2019.

In Table 1, it can be seen that GDP per capita and the number of beds per 1,000 population in hospitals have a negative correlation with the mortality rate, i.e., the higher the GDP per capita and the more beds per 1,000 population in hospitals, the lower the mortality rate, which is consistent with the findings of Ying et al. [10]. The reason for this situation is that economically developed areas have better basic medical security, medical institutions have more beds per thousand people, the social medical system is more perfect, and people's ability to resist social risks is stronger, so the mortality rate will decrease.

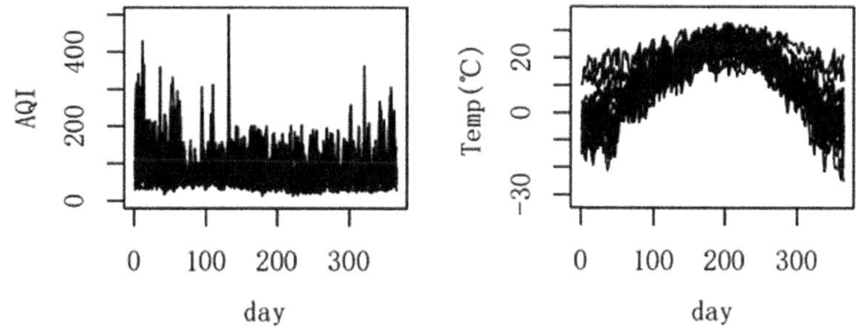

Fig. 1 Air quality index and temperature in some cities

Table 1 The estimated value of regression coefficient and its significance level

$\hat{\gamma}$	Estimate	Std. error	t value	Pr(>|t|)
$\hat{\gamma}_{GDP}$	−1.126e−05	4.007e−06	−2.810	0.006311 **
$\hat{\gamma}_{Number}$	−2.062e−01	1.920e−01	−2.654	0.0378 *

The results of the regression coefficient functions $\hat{\beta}(t_1)$ and $\hat{\beta}(t_2)$ for the function-based predictor variables are shown in Fig. 2. It can be seen that the air quality index and population mortality have a significant positive correlation, i.e. the higher the air quality index, the greater the risk of disease and the higher the population mortality rate, which is consistent with the research results of Yuan et al. [11]. The relationship between temperature and population mortality can be seen in the graph, which shows a positive correlation between temperature and population mortality in general, with the exception of spring. The higher the temperature, the faster the disease multiplies and spreads, which greatly increases the risk of disease and therefore the mortality rate. The warmer the weather in spring, the more the pores of the skin open, and the lower the temperature, the more the cold wind and the epidemic can attack the body and damage the health of the population, thus a negative correlation between temperature and mortality in spring. Figures 3 and 4 show the variation characteristics of $\hat{\beta}(t_1, t_2)$ within the confidence interval. As can be seen from the figure, when $t_2 \in [-30, -10]$, $\hat{\beta}(t_1, t_2)$ first decreases, then increases, then decreases and finally decreases with the increase of t_1; when $t_2 \in [-10, 30]$, $\hat{\beta}(t_1, t_2)$ first increases, then decreases, then increases and then decreases with the increase of t_1.

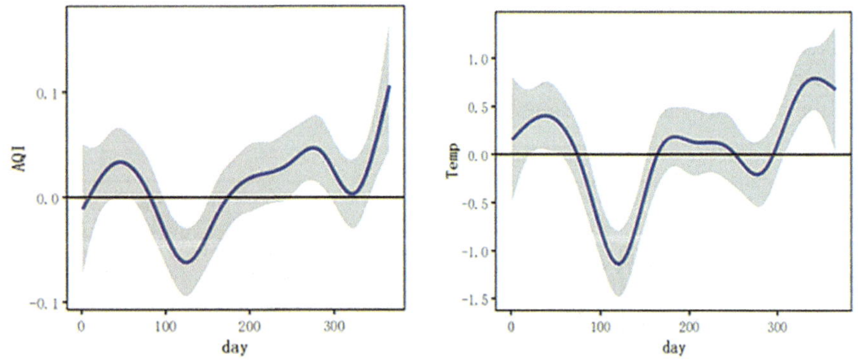

Fig. 2 $\hat{\beta}_j(t_j)$ and its 95% confidence interval band

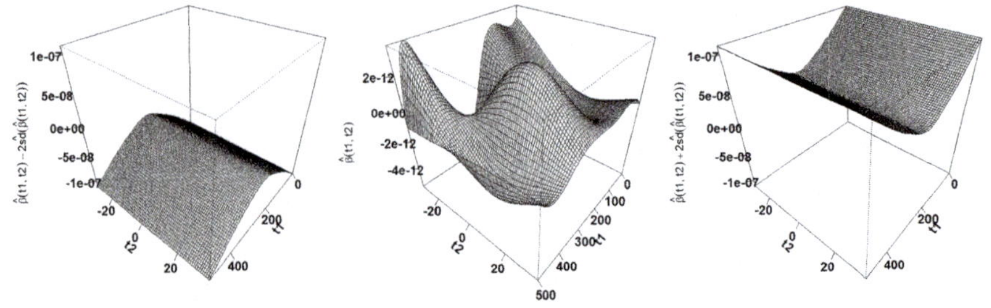

Fig. 3 Visual three-dimensional diagram of $\hat{\beta}(t_1, t_2)$

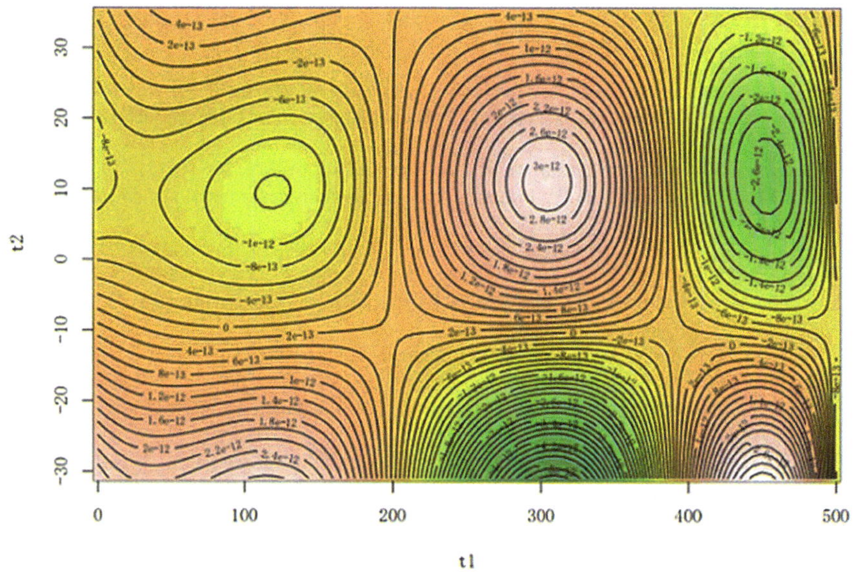

Fig. 4 Contour map of $\hat{\beta}(t_1, t_2)$

4 Conclusion

This paper has presented a generalized partially functional linear model with interaction terms that considers both the effects of functional and scalar predictor variables and the interactions between functional predictor variables. It has been applied to investigate the factors influencing population mortality. The results have shown that there has been an overall positive correlation between the air quality index and temperature and the population mortality rate. Meanwhile, the effects of GDP per capita and the number of beds per 1,000 population in hospitals on the population mortality rate have both been negatively correlated. The air quality index and temperature have been two factors that have affected each other, therefore, considering the interaction of the two in the study of population mortality has made the prediction results more scientifically accurate.

Population mortality is the result of the combined effect of the social and natural environment, and population mortality is an important indicator of the extent of population mortality. Therefore, in order to curb the increase in population mortality, it is all the more important for us to actively deal with the existing environmental problems, upgrade the medical level, improve the medical environment and provide more convenient basic medical facilities for the people.

Acknowledgements Yujie Talent Project of North China University of Technology, No. 107051360023XN075-04.

References

1. James, J.M.: Generalized linear models with functional predictors. J. R. Stat. Soc. **64**(3), (2002)
2. Locantore, N., Marron, J.S.: Robust principal component analysis for functional data. Test **8**(1), (1998)
3. Ramsay, J.O.: When the data are functions. Psychom. **47**(4), 379–396 (1982)
4. Ramsay, J.O., Silverman, B.W.: Principal components analysis for functional data. Funct. Data Anal. (1997)
5. Nelder, J.A., Wedderburn, R.W.M.: Generalized linear models. J. Roy. Stat. Soc. **135**(3), 370–384 (1972)
6. Müller, H.G., Stadtmüller, U.: Generalized functional linear models. Ann. Stat. **33**(2), 774–805 (2005)
7. Yang, W.H., Wikle, C.K., Holan, S.H., Wildhaber, M.L.: Ecological prediction with nonlinear multivariate Time-Frequency functional data models. J. Agric., Biol., Environ. Stat. (2013)
8. Usset, J., Staicu, A.M., Maity, A.: Interaction models for functional regression. Comput. Stat. Data Anal. **94**, 317–329 (2016)
9. Kong, D.H.: Partially functional linear regression in high dimensions. Biom. **139**(10), 147–159 (2016)
10. Kui, Y., Xudong, L.: The cvolution of spatial pattern and innuencing factors of population mortality in china. World Reg. Stud. **31**(2), 440–452 (2022)
11. Jing, Y., Xu, C.Z.: The influence study on air pollution and economic development level to population mortality: the empirical analysis based on vector autoregressive model. Ecol. Econ. **34**(3), 156–159 (2018)

Research on Sleep Quality Based on Generalized Partially Function Linear Model with Unknown Link Function

Songxuan Li, Kejing Mao, and Weiwei Xiao

Abstract This paper investigates the content related to sleep quality. A generalized partially functional linear model in which the response variable is scalar and the predictor variable is both functional and scalar is constructed by using the sleep quality of 22 volunteers as the response variable, the hourly heart rate as a functional predictor variable, and Wake After Sleep Onset, the Number of Awakenings during the night, and Daily stress Inventory value as scalar predictor variables, where the model exact link function was not specified. Finally, it was concluded that HR was positively correlated with sleep quality from 9 am to about 15 pm and negatively correlated at other times, and that the WASO, Number, and DSI were inversely correlated with sleep quality, which was also largely in line with our expectations.

Keywords Generalized partially function linear · Functional principal component analysis · Sleep quality

1 Introduction

In 1982, Ramsay [1] introduced functional data, which provided a solid foundation for the development of functional data analysis. The functional generalized linear regression model is an extension of the functional data linear regression model. The generalized linear model can be used to describe both continuous and discrete variables, which makes it more widely applicable in fields such as medicine, biology, and social sciences for describing attribute data and count data. James [2] incorporated functional predictors into the generalized linear model. Müller et al. [3] proposed an asymptotic theory for the generalized functional regression model, which obtained the asymptotic normality of the model parameters and the confidence intervals through the generalized covariance operator. A rough regularization

S. Li · K. Mao · W. Xiao (✉)
North China University of Technology, Beijing, China
e-mail: xiaoww@ncut.edu.cn

W. Wang et al. (eds.), *Computational Statistics and Data Intelligence*, Springer Proceedings in Mathematics & Statistics 463,
https://doi.org/10.1007/978-981-97-4438-1_5

method was proposed by Shang and Cheng [4] for performing nonparametric inference in the generalized functional linear model. Wong et al. [5] studied parametric and nonparametric estimation of multiple predictor variables to predict scalar response variables.

Sleep is essential for maintaining normal physiological function in the human body. Adequate sleep can promote body recovery and repair, as well as help maintain physical and mental health. Buysse et al. [6] discusses the Pittsburgh Sleep Quality Index (PSQI), which is a questionnaire designed to assess sleep quality and disturbances over a one-month time interval. Curcio et al. [7] aimed to assess the validity of the PSQI in the Italian population and found that certain aspects of the PSQI questionnaire scores were associated with factors such as gender and age. Vitale et al. [8] conducted a study on the influence of chronotype on activity circadian rhythm and sleep, as well as the differences in sleep quality between weekdays and weekends. Chalmers [9] explores the associations between sleep quality and heart rate variability, and their implications for a biological model of stress detection using wearable technology. Geng et al. [10] describes a personalized method for recognizing wake and sleep states based on the combined shapelets and K-means algorithm.

Based on previous research, this article develops a generalized partially functional linear model with an unknown link function. Applying this model, the study investigates the factors affecting sleep quality from four aspects: researchers' HR, the WASO, the Number of Awakenings, and the DSI. This can help people maintain a good sleep state in their daily lives, thereby promoting physical and mental health.

2 Generalized Partially Function Linear Models

We observe independent and identically distributed data.

$$(Y_i, X_{ij}(t_j), Z_i), i = 1, \ldots, n, j = 1, \ldots, d$$

where Y_i is the response variable, $X_{ij}(t_j)$ is a random curve corresponding to a square integrable stochastic process on the real interval t, which is called the functional predictor. $Z_i = (Z_1, Z_2, \ldots, Z_q)$ is a q dimensional random vector, which is called the scalar predictor[11]. A model is established to describe the relationship between $Y_i, X_{ij}(t_j)$ and Z_i

$$Y_i = g\left(\sum_{j=1}^{d} \int_{T_j} X_{ij}(t_j)\beta_j(t_j)dt_j + Z_i\gamma\right) + \varepsilon_i \tag{1}$$

where $\beta_j(\cdot)$ represents the regression coefficients of $X_{ij}(t_j)$, and vector $\gamma = (\gamma_1, \gamma_2, \ldots, \gamma_q)^T$ represents the regression coefficients of Z, both of which need to be estimated. ε is an independent random error variable that satisfies $E[\varepsilon \,|X_j(t_j), Z] = 0$.

3 Application to Sleep Quality

The data in this article comes from the Multi-Modal Ambulatory Sleep and Health Monitoring (MMASH) dataset of healthy individuals' activity and sleep, provided by the PhysioNet database, which was collected and provided through a collaboration between BioBeats and researchers from the University of Pisa [12].

Our aim is to investigate the hourly Heart Rate (HR), Wake After Sleep Onset (WASO), Number of Awakenings during the night (Number) and Daily Stress Inventory value (DSI) on sleep quality in 21 volunteers.

The response variable (PSQI) was obtained from the Pittsburgh Sleep Quality Index questionnaire [7]. Based on this criterion, we categorized the sleep quality of the 21 volunteers. A score less than 6 was coded as 1 denoting good sleep quality, while a score greater than 6 was coded as 0 indicating poor sleep quality. This was done to ensure that the response variable follows a Bernoulli distribution. In our data study, 6 participants had poor sleep quality, while 15 had good sleep quality. The functional predictor variable is the hourly heart rate of 21 volunteers, shown as the heart rate changes per hour for each participant in Fig. 1. In processing the data, the first step is to center the data and then select the functional principal components with a cumulative contribution rate of 85%. In this case, we choose as the number of truncated orthogonal bases for the standard basis. The scalar predictor variables are WASO, The Number of Awakenings and DSI. DSI is a self-reported measure consisting of 58 items that assess daily stress, where the higher the daily stress score, the higher the frequency and extent of events and perceived daily stress.

For better sleep quality, it is important to maintain a lower heart rate during sleep. This can help our body enter deeper sleep. When the heart rate is too high, it indicates that the body's nervous system is in an excited state, which may make it difficult to fall asleep. On the other hand, when the heart rate is too low, it may cause a series of problems such as poor blood circulation, which can also affect the quality of sleep. Therefore, maintaining an appropriate heart rate level is very important for promoting good sleep quality. In Fig. 2, the red line represents the estimated regression coefficient $\hat{\beta}(t)$, and the gray area represents its 95% confidence interval. By observing the trend of the curve, it can be seen that from around 9 am to 3 pm, heart rate and sleep quality have a positive correlation, while at other times they show a negative correlation. This period happens to be the most focused time for our work, so we can increase some exercise during this time to relax and improve our sleep quality at night.

The sleep cycle is typically divided into three stages: the initial sleep stage, the light sleep stage, and the deep sleep stage. During the initial and light sleep stages, our body gradually relaxes, and our breathing and heart rate begin to slow down. However, our sleep state is not yet stable, and we are more easily awakened by external factors. During the deep sleep stage, the body enters the deepest state of rest and enters a state that is more difficult to be awakened from.

Table 1 shows the scalar regression coefficient $\hat{\gamma}$.By observing Table 1, we can see that all three scalar predictive variables meet the criterion of $P < 0.05$.

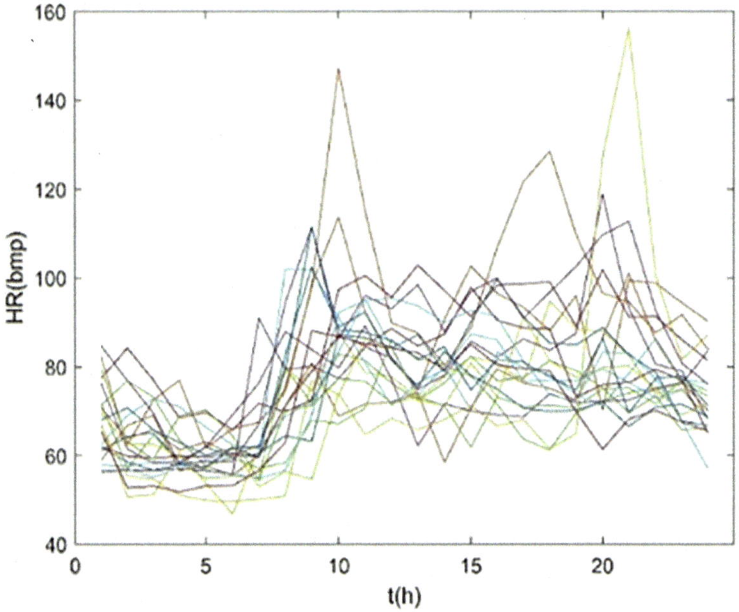

Fig. 1 The heart rate changes per hour for each participant

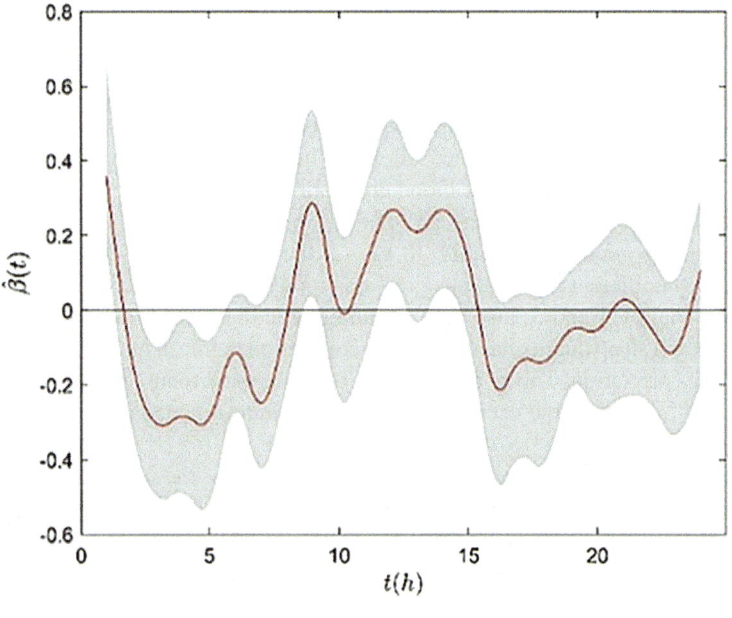

Fig. 2 Regression parameters $\hat{\beta}(t)$ and its 95% confidence intervals

Table 1 The scalar regression coefficient $\hat{\gamma}$

	Estimate	Std. Error	t value	Pr(>|t|)	
$\hat{\gamma}_{WASO}$	−0.3488	0.1632	−2.1373	0.04836	*
$\hat{\gamma}_{Number}$	−0.9316	0.7260	−1.2832	0.02177	*
$\hat{\gamma}_{DSI}$	−0.1026	0.1806	−0.5681	0.04778	*

WASO is inversely proportional to sleep quality. A longer time from falling asleep to the first awakening means that we spend more time in the initial and light sleep stages. But good sleep quality usually refers to spending a longer time in the deep sleep stage. Therefore, the longer the time from falling asleep to the first awakening, the worse the sleep quality is. The Number of Awakenings is also inversely proportional to sleep quality. When we wake up frequently at night, we need to spend more time falling asleep, which disrupts our sleep cycle and affects our sleep quality. DSI is also inversely proportional to sleep quality. Long-term stress can lead to physical problems, and these physical signals may further lead to insomnia, forming a vicious cycle. Therefore, timely stress relief is also important for us.

4 Conclusion

A generalized partially functional linear model with an unknown link function is applied to the study of sleep quality, and it is found that the correlation between heart rate and sleep quality is positive in the morning from 9 am to around 3 pm, and negative at other times. Additionally, the WASO, the Number of Awakenings, and the DSI are all inversely proportional to sleep quality. These results can help people better understand the influencing factors of sleep quality and take corresponding measures, such as adding appropriate exercise, maintaining a good heart rate, matching diets reasonably, learning to relieve stress in time, and establishing healthy sleep habits to improve sleep quality. However, there are still issues that need to be further considered in the actual application of this model, such as the lack of consideration of the interaction between predictive variables, which can be further explored in subsequent research.

Acknowledgements Yujie Talent Project of North China University of Technology, No. 107051360023XN075-04.

References

1. Ramsay, J.O.: When the data are functions. Psychom. **47**(4), 379–396 (1982)
2. James, G.M.: Generalized linear models with functional predictors. J. Roy. Stat. Soc. B **64**(3), 411–432 (2002)
3. Müller, H.G., Stadt Müller, U.: Generalized functional linear models. Ann. Stat. **33**(2), 774–805 (2005)
4. Shang, Z., Cheng, G.: Nonparametric inference in generalized functional linear models. Ann. Stat. **43**(4), 1742–1773 (2015)
5. Wong, R.K.W., Li, Y., Zhu, Z.: Partially linear functional additive models for multivariate functional data. J. Am. Statal Assoc. **114**(525), 406–418 (2018)
6. Buysse, D.J., Reynolds, C.F., Monk, T.H., Berman, S.R., Kupfer, D.J.: The Pittsburgh sleep quality index: A new instrument for psychiatric practice and research. Psychiatry Res. **28**(2), 193–213 (1989)
7. Curcio, G., Tempesta, D., Scarlata, S., et al.: Validity of the Italian version of the Pittsburgh Sleep Quality Index (PSQI). Neurol. Sci. **34**, 511–519 (2013)
8. Vitale, J.A., Roveda, E., Montaruli, A., et al.: Chronotype influences activity circadian rhythm and sleep: Differences in sleep quality between weekdays and weekend. Chronobiol. Int. **32**(3), 405–415 (2014)
9. Chalmers, T., Hickey, B.A., Newton, P., et al.: Associations between sleep quality and heartrate variability; implications for a biological model of stress detection using wearable technology. Public Health **19**(9), 5770 (2022)
10. Geng, D., Qin, Z., Wang, J., et al.: Personalized recognition of wake/sleep state based on the combined shapelets and K-means algorithm. Biomed. Signal Process. Control **17**, 103132 (2022)
11. Xiao, W., Wang, Y., Liu, H.: Generalized partially functional linear model. Sci. Rep. **11**, 23428 (2021)
12. Rossi, A., Da Pozzo, E., Menicagli, D., et al.: A public dataset of 24-h Multi-Levels Psycho-Physiological responses in young healthy adults. Data **5**(4), 91 (2020)

Decompositions and Packing of Complete 3-Uniform Hypergraph Into Hyperstar of a Certain Type

Hongtao Zhao and Yuzhe Zhu

Abstract Paola Bonacini[6] defined the hyperstar which is a hypergraph with m 3-edges such that all of them have two vertices in common and is denoted by $S^{(3)}(2, m+2)$. In this paper, we consider the $S^{(3)}(2, 6)$ decomposition and packing of complete 3-uniform hypergraph. Firstly, the necessary condition for the decomposition of the complete 3-uniform hypergraph into an $S^{(3)}(2, 6)$ is $v = 0, 1, 2, 4, 6 \pmod 8$, $v \geq 6$. Secondly, according to the recursive constructions, the required designs of small orders are found. For the hypergraphs with large orders, they can be recursively generated by some designs of small orders. Then, it is proved that the above necessary conditions are sufficient except $v = 6$, that is, there exists an $S^{(3)}(2, 6)$ decomposition of a complete 3-uniform hypergraph if and only if $v = 0, 1, 2, 4, 6 \pmod 8$, $v > 6$. Finally, we prove that there exists a maximum $S^{(3)}(2, 6)-$ packing, except $v = 6, 7$. And we find the exact value of the packing number $d(t, T, v)$ for any integer $v \geq 6$.

Keywords Complete t-Uniform hypergraph · (H, Γ)-Design · Packing design · Maximum packing · Hyperstar

1 Introduction

A hypergraph H consists of a finite nonempty set and its subset family. The elements of the set V are called vertices, and the elements of the set E are called hyperedges. If for each $e \in E$, we have $|e| = t$, then H is said to be $t-$ uniform. In this case, if the edge set E contains each $t-$ subset of V exactly once and $|V| = v$, the hypergraph is called a complete $t-$ uniform hypergraph $K_v^{(3)}$ of order v.

Let H be a complete $t-$ uniform hypergraph and Γ be a set of $t-$ uniform hypergraphs. The $\Gamma-$ decomposition of a hypergraph H is to decompose the edge of H into a set of sub-hypergraphs, and each sub-hypergraph is isomorphic to a hypergraph in

H. Zhao · Y. Zhu (✉)
North China Electric Power University, Beinong Road No.2, Beijing, China
e-mail: zyz881234560@163.com

W. Wang et al. (eds.), *Computational Statistics and Data Intelligence*, Springer Proceedings in Mathematics & Statistics 463,
https://doi.org/10.1007/978-981-97-4438-1_6

Γ. The $\Gamma-$ decomposition of a hypergraph H is called a $(H, \Gamma)-$ design. $(H, \Gamma)-$ design is also denoted a $t-$ balanced design $S(t, \Gamma, v)$.

Let H be a complete $t-$ uniform hypergraph, $H = K_v^{(3)}$, then $(H, \Gamma)-$ packing is a binary $(X, \mathcal{B})$, where X is the finite vertex set of H, $\mathcal{B}$ is the set of some sub-hypergraphs of H, each hypergraph of $\mathcal{B}$ is isomorphic to a hypergraph in Γ, called a block, and each edge of H is contained in at most one block of $\mathcal{B}$. Decomposition design is a special kind of packing design. In this case, a $(H, \Gamma)-$ packing design can be denoted by $P(t, \Gamma, v)$.

When there is no packing $(X, \mathcal{A})$ such that $|\mathcal{B}| < |\mathcal{A}|$, then packing $(X, \mathcal{B})$ is called the maximum packing, denoted by $MP(t, \Gamma, v)$. The number of blocks in the maximum packing $MP(t, \Gamma, v)$ is called the packing number of the packing, denoted by $d(t, \Gamma, v)$.

The problem of hyperstar was first proposed in 1987. Lon Zbigniew [1] defined the hyperstar which is a 3-uniform hypergraph consisting of one vertex of degree m and $2m$ vertices of degree one, and is denoted by $S_m^{(3)}$. Necessary and sufficient conditions for the existence of $S_m^{(3)}-$ decompositions of $K_v^{(3)}$ are given in reference [2], for $m \in \{4, 5, 6\}$ and settled in reference [3] for any m. The maximum $S_m^{(3)}-$ packings of $K_v^{(3)}$ are given in reference [4]. In 2014, Hoffman solved the packing and covering problems for any $k-$ star [5]. Paola Bonacini [6] defined a new hyperstar which is a hypergraph with m 3-edges such that all of them have two vertices in common and is denoted by $S^{(3)}(2, m + 2)$.

Several authors have given many results on the $\Gamma-$ decomposition, packing and covering problem of 3-uniform hypergraphs, Γ is $K_4^{(3)} - e$, $W_4^{(3)}$, and $K_4^{(3)} + e$ [7–12]. Amber Armstrong [13] found maximum packings of $\lambda K_v^{(3)}$ with copies of the symmetric triple-hyperstar with 4 edges.

In this paper, for the first time, we consider decompositions and packing of complete 3-uniform hypergraph into $S^{(3)}(2, 6)$ hyperstar. For convenience, we always use T in place of $S^{(3)}(2, 6)$. Based on calculations, we can get about the upper bound of the packing number: $d(t, T, v) \leq \left\lfloor \frac{v(v-1)(v-2)}{24} \right\rfloor$. In this paper, we show that there exists a $T-$ decomposition of $K_v^{(3)}$ if and only if $v = 0, 1, 2, 4, 6 \pmod 8$, $v > 6$. For any integer $v \geq 6$, except $v = 6, 7$, there exists $MP(3, T, v)$, and $d(t, T, v) = \left\lfloor \frac{v(v-1)(v-2)}{24} \right\rfloor$. Especially, $d(3, T, 6) = 3$, $d(3, T, 7) = 7$.

2 Recursive Constructions

Let $s(\leq v)$ be a nonnegative integer. A $(K_v^{(t)} \backslash K_s^{(t)}, \Gamma)-$ design is an ordered triple $(X, Y, \mathcal{B})$ where X is the vertex set of $K_v^{(t)} \backslash K_s^{(t)}$, Y is the vertex set of $K_s^{(t)}$, Y is called a hole, $\mathcal{B}$ is the set of sub-hypergraphs of $K_v^{(t)} \backslash K_s^{(t)}$, each sub-hypergraph (called a block) is isomorphic to a hypergraph in Γ and each edge of $K_v^{(t)} \backslash K_s^{(t)}$ exactly contains a block of $\mathcal{B}$, which is denoted by $HS(t, \Gamma; v, s)$.

We use the usual exponential notation for the types of groups divisible $(H, \Gamma)-$ designs. Then type $g_1^{a_1} g_2^{a_2} \cdots g_m^{a_m}$ denotes that there are a_i groups of size g_i, $1 \leq$

$i \leq m$. A group divisible $(H, \Gamma)-$ design of type $g_1^{a_1} g_2^{a_2} \cdots g_m^{a_m}$ can be denoted by GDD(t, Γ, v), where $v = \sum_{i=1}^{m} a_i g_i$. If Γ contains only one hypergraph J, we write GDD$(t, \{J\}, v)$ as GDD(t, J, v).

Let $V = V_1 \cup V_2 \cup V_3$. The complete tripartite 3-uniform hypergraph H consists the vertex set V and hyperedge set E, where $V = V_1 \cup V_2 \cup V_3$, $V_i \cap V_j = \emptyset$ and $E = \{e | e \subseteq V, |e| = 3, e \cap V_i \neq \emptyset\}$ for $1 \leq i < j \leq 3$. The hypergraph is denoted by $K^{(3)}_{v_1,v_2,v_3}$.

Let $V = V_1 \cup V_2$. The complete bipartite 3-uniform hypergraph H consists the vertex set V and hyperedge set E, where $V = V_1 \cup V_2$, $V_1 \cap V_2 = \emptyset$ and $E = \{e | e \subseteq V, |e| = 3, e \cap V_i \neq \emptyset\}$ for $i = 1, 2$. The hypergraph is denoted by $L^{(3)}_{v_1,v_2}$.

The necessary conditions for the existence of an $S(t, \Gamma, v)$ are as follows:

Lemma 1. (11) Let Γ be any set of 3-uniform hypergraphs and $J \in \Gamma$. Let $d^*(x_1, x_2)$ be the number of edges in J containing the two vertices x_1 and x_2. The following are necessary conditions for the existence of an $S(t, \Gamma, v)$:

(1) $v \geq \min\{|V(J)| : J \in \Gamma\}$;

(2) $\binom{v}{3} \equiv 0, (\bmod\ d_0)$, where $d_0 = \gcd\{|E(J)| : j \in \Gamma\}$;

(3) $\binom{v-1}{2} \equiv 0, (\bmod\ d_1)$, where $d_1 = \gcd\{d(x) : x \in V(J), J \in \Gamma\}$;

(4) $\binom{v-2}{1} \equiv 0, (\bmod\, d_2)$, where $d_2 = \gcd\{d_J^*(x_1, x_2) : x_1 \neq x_2, x_1, x_2 \in V(J), J \in \Gamma\}$.

Lemma 2. Let a, r and l be nonnegative integers. Let $v = ar + l$, and $v \geq 3$. Under the given conditions, there exists a decomposition of $K_v^{(3)}$, which is comprised of isomorphic copies of each of the following:

(1) $K_l^{(3)}$ if $r = 0$
(2) $K_{a+l}^{(3)}$ if $r \geq 1$
(3) $K_{a+l}^{(3)} \backslash K_l^{(3)}$ if $r \geq 2$
(4) $K_{l,a,a}^{(3)} \cup L_{a,a}^{(3)}$ if $r \geq 2$
(5) $K_{a,a,a}^{(3)}$ if $r \geq 3$

Proof. When $r \in \{0, 1\}$, the decomposition is trivial. Correspondingly, if $a = 0$ and $l \geq 3$, then $K_l^{(3)} = K_{a+l}^{(3)} = K_{ar+l}^{(3)}$, and $K_{a+l}^{(3)} \backslash K_l^{(3)}$, $K_{l,a,a}^{(3)} \cup L_{a,a}^{(3)}$, $K_{a,a,a}^{(3)}$ are empty (i.e., without any edges), and the decomposition is also trivial.

Let $A_0, A_1, A_2, \cdots, A_r$ be pairwise-disjoint sets of vertices with $|A_0| = l$, $|A_1| = |A_2| = \cdots = |A_r| = a$. Next, we suppose that $r \geq 2$ and $a \geq 1$.

Let $V = A_0 \cup A_1 \cup A_2 \cup \cdots \cup A_r$, then the decomposition of $K_v^{(3)}$ can be viewed as the (edge-disjoint) union.

$$K^{(3)}_{A_1 \cup A_0} \cup \bigcup_{2 \leq i \leq r} (K^{(3)}_{A_i \cup A_0} \backslash K^{(3)}_{A_0}) \cup \bigcup_{1 \leq i \leq j \leq r} (K^{(3)}_{A_0,A_i,A_j} \cup L^{(3)}_{A_i,A_j}) \cup \bigcup_{1 \leq i < j < k \leq r} (K^{(3)}_{A_i,A_j,A_k}).$$

3 Some Small Orders

We use (a, b, c, d, e, f) to denote any hypergraph isomorphic to the $S^{(3)}(2, 6)$ with vertex set $\{a, b, c, d, e, f\}$ and edge set $\{a, b, c\}$, $\{a, b, d\}$, $\{a, b, e\}$, $\{a, b, f\}$ which a, b is the center as seen in Fig. 1. Let $\mathbb{Z}_m$ denote the group of integers modulo m. Note that Repeated orbits are not counted.

Next, we give some examples of $T-$ decompositions to prove our main result.

Example 1. Let $V(K_8^{(3)}) = \mathbb{Z}_8$, then a $T-$ decomposition of $K_8^{(3)}$ consists of the blocks in B:

$$B = (0, 1, 2, 3, 4, 5), (2, 3, 1, 5, 7, 6), (0, 6, 1, 5, 2, 3), (0, 2, 4, 5, 7, 3), (2, 4, 1, 2, 7, 6),$$
$$(2, 5, 4, 7, 1, 6), (1, 6, 2, 3, 5, 7), (1, 7, 2, 5, 3, 4), (3, 5, 1, 0, 7, 6), (3, 4, 5, 0, 7, 1),$$
$$(4, 5, 6, 7, 0, 1), (4, 6, 1, 3, 0, 4), (6, 7, 2, 5, 0, 3), (0, 7, 1, 4, 3, 5).$$

Example 2. Let $V(K_9^{(3)}) = \mathbb{Z}_9$, then a $T-$ decomposition of $K_9^{(3)}$ consists of the blocks in B:

$$B = (1, 2, 0, 3, 7, 8), (0, 3, 1, 2, 4, 6), (1, 4, 0, 2, 6, 8), (1, 5, 0, 4, 3, 7), (0, 6, 1, 2, 5, 7),$$
$$(2, 4, 0, 3, 6, 5), (0, 5, 2, 3, 4, 8), (2, 5, 1, 7, 3, 6), (1, 6, 3, 5, 2, 8), (3, 6, 2, 4, 5, 7),$$
$$(2, 7, 3, 6, 0, 4), (2, 8, 3, 7, 0, 5), (4, 8, 7, 2, 5, 0), (4, 3, 8, 7, 1, 5), (5, 7, 3, 4, 6, 8),$$
$$(3, 8, 5, 7, 6, 0), (4, 6, 5, 8, 0, 7), (6, 8, 7, 2, 5, 0), (0, 7, 5, 3, 8, 4), (1, 8, 7, 0, 3, 5),$$
$$(1, 7, 0, 6, 4, 3).$$

Example 3. Let $V(K_{10}^{(3)}) = \mathbb{Z}_{10}$. Under the action of $\mathbb{Z}_{10}$, all 120 edges of $K_{10}^{(3)}$ can be divided into 12 orbit representatives. Base blocks are given as follows:

$$B = (0, 3, 4, 5, 8, 9), (0, 4, 2, 5, 7, 9), (1, 9, 0, 2, 5, 8).$$

It's easy to check that $T-$ blocks in B cover all the 12 orbit representatives, and will generate 30 blocks by $+1$ modulo 10.

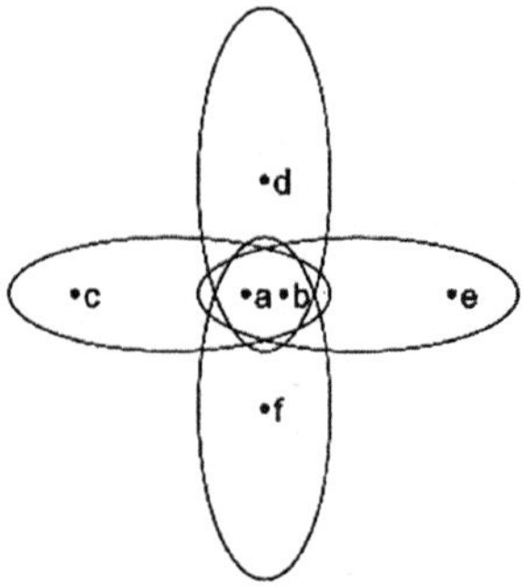

Fig. 1 Hyperstar $S^{(3)}(2, 6)$ denoted by (a, b, c, d, e, f)

Example 4. Let $V(K_{12}^{(3)}) = \mathbb{Z}_{11} \cup \{\infty\}$. Under the action of $\mathbb{Z}_{11}$, all 220 edges of $K_{12}^{(3)}$ can be divided into 20 orbit representatives. Base blocks are given as follows:

$$B = (0, 1, \infty, 2, 4, 6), (1, 4, 5, 10, 7, \infty),$$
$$(3, 5, 2, \infty, 6, 10), (0, 7, \infty, 4, 6, 9), (4, 10, 2, 3, \infty, 6).$$

Then $T-$ blocks in B cover all the 20 orbit representatives and will generate 55 blocks by $+1$ modulo 11.

Example 5. Let $V(K_{14}^{(3)}) = \mathbb{Z}_{14}$. Under the action of $\mathbb{Z}_{14}$, all 364 edges of $K_{14}^{(3)}$ can be divided into 26 orbit representatives. Base blocks are given as follows:

$$B = (2, 7, 1, 10, 9, 5), (0, 5, 4, 1, 7, 9), (0, 2, 1, 12, 3, 13), (5, 13, 7, 6, 9, 8),$$
$$(4, 10, 6, 9, 7, 8), (0, 3, 4, 13, 10, 12), (1, 8, 2, 11, 4, 9).$$

Then $T-$ blocks in B cover all the 26 orbit representatives and will generate 91 blocks by $+1$ modulo 14.

Example 6. Let $V(K_{16}^{(3)}) = \mathbb{Z}_{14} \cup \{\infty_1, \infty_2\}$. Under the action of $\mathbb{Z}_{14}$, all 560 edges of $K_{14}^{(3)}$ can be divided into 40 orbit representatives. Base blocks are given as follows:

$$B = (0, 1, \infty_1, \infty_2, 2, 9), (0, 10, 1, 11, 13, 12), (0, 2, 3, 13, 7, \infty_1), (0, 10, 2, \infty_1, 9, 6),$$
$$(0, \infty_2, 4, \infty_1, 5, 6), (1, 4, \infty_1, 9, \infty_2, 10), (3, 11, 4, 0, \infty_1, 7), (0, 12, \infty_2, 6, 3, 7),$$
$$(1, 8, 5, \infty_2, 9, 12), (0, 7, 1, 3, \infty_1, 10), (4, 9, 7, 0, 3, \infty_1).$$

Then $T-$ blocks in B cover all the 40 orbit representatives and will generate 140 blocks by $+1$ modulo 14.

Example 7. Let $V(L_{8,8}^{(3)}) = \mathbb{Z}_{16}$, with vertex partition $\{\{0, 2, 4, 6, 8, 10, 12, 14\}, \{1, 3, 5, 7, 9, 11, 13, 15\}\}$. Under the action of $\mathbb{Z}_{16}$, all 448 edges of $L_{8,8}^{(3)}$ can be divided into 28 orbit representatives. Base blocks are given as follows:

$$B = (11, 14, 12, 15, 1, 5), (0, 5, 15, 6, 7, 10), (0, 1, 7, 12, 9, 13), (4, 5, 9, 6, 12, 14),$$
$$(8, 10, 11, 13, 1, 15), (3, 7, 4, 0, 12, 14), (0, 13, 5, 9, 6, 8).$$

Then $T-$ blocks in B cover all the 28 orbit representatives and will generate 112 blocks by $+1$ modulo 16.

Example 8. Let $V(K_{1,8,8}^{(3)} \cup L_{8,8}^{(3)}) = \mathbb{Z}_{16} \cup \{\infty\}$, with vertex partition $\{\{\infty\}, \{0, 2, 4, 6, 8, 10, 12, 14\}, \{1, 3, 5, 7, 9, 11, 13, 15\}\}$. Under the action of $\mathbb{Z}_{16}$, all 512 edges of $K_{1,8,8}^{(3)} \cup L_{8,8}^{(3)}$ can be divided into 32 orbit representatives. Base blocks are given as follows:

$$B = (0, 7, 1, 14, 6, 12), (0, 14, 15, 9, 1, 11), (0, 4, 1, 15, 13, 5), (0, 3, 1, 6, 4, 10),$$
$$(1, 6, 0, 3, 13, 11), (2, 13, 11, 5, 3, 10), (1, 8, 11, 0, 5, 9), (0, \infty, 1, 3, 5, 7).$$

Then $T-$ blocks in B cover all the 32 orbit representatives and will generate 128 blocks by $+1$ modulo 16, where $\infty + 1 = \infty$.

Example 9. Let $V(K^{(3)}_{2,8,8} \cup L^{(3)}_{8,8}) = \mathbb{Z}_{16} \cup \{\infty_1, \infty_2\}$, with vertex partition $\{\{\infty_1, \infty_2\}, \{0, 2, 4, 6, 8, 10, 12, 14\}, \{1, 3, 5, 7, 9, 11, 13, 15\}\}$. Under the action of $\mathbb{Z}_{16}$, all 576 edges of $K^{(3)}_{2,8,8} \cup L^{(3)}_{8,8}$ can be divided into 36 orbit representatives. Base blocks are given as follows:

$$\begin{aligned} B = {} & (3, 7, 2, 12, 0, 14), (0, 6, 9, 1, 3, 15), (8, 15, 12, 1, 13, 14), (0, 3, 2, \infty_2, 15, \infty_1), \\ & (0, 1, \infty_2, 2, \infty_1, 8), (2, 5, 13, 0, 6, 3), (1, 10, 2, \infty_1, 4, \infty_2), (4, 9, 7, 4, 1, 14), \\ & (1, 6, \infty_1, 7, 15, \infty_2). \end{aligned}$$

Then $T-$ blocks in B cover all the 36 orbit representatives, and will generate 144 blocks by $+1$ modulo 16, where $\infty_j + 1 = \infty_j$ for $j \in \{1, 2\}$.

Example 10. Let $V(K^{(3)}_{3,8,8}) = \mathbb{Z}_{16} \cup \{\infty_1, \infty_2, \infty_3\}$, with vertex partition $\{\{\infty_1, \infty_2, \infty_3\}, \{0, 2, 4, 6, 8, 10, 12, 14\}, \{1, 3, 5, 7, 9, 11, 13, 15\}\}$. Under the action of $\mathbb{Z}_{16}$, all 192 edges of $K^{(3)}_{3,8,8}$ can be divided into 12 orbit representatives. Base blocks are given as follows:

$$B = (3, \infty_1, 0, 12, 8, 4), (3, \infty_2, 4, 6, 14, 12), (\infty_3, 8, 7, 5, 3, 1).$$

Then $T-$ blocks in B cover all the 12 orbit representatives, and will generate 48 blocks by $+1$ modulo 16, where $\infty_j + 1 = \infty_j$ for $j \in \{1, 2, 3\}$.

Example 11. Let $V(K^{(3)}_{4,8,8}) = \mathbb{Z}_{16} \cup \{\infty_1, \infty_2, \infty_3, \infty_4\}$, with vertex partition $\{\{\infty_1, \infty_2, \infty_3, \infty_4\}, \{0, 2, 4, 6, 8, 10, 12, 14\}, \{1, 3, 5, 7, 9, 11, 13, 15\}\}$. Under the action of $\mathbb{Z}_{16}$, all 256 edges of $K^{(3)}_{4,8,8}$ can be divided into 16 orbit representatives. Base blocks are given as follows:

$$B = (0, \infty_1, 15, 11, 7, 3), (2, \infty_2, 1, 9, 5, 13), (7, \infty_3, 8, 14, 2, 10), (5, \infty_4, 6, 12, 0, 8).$$

Then $T-$ blocks in B cover all the 16 orbit representatives, and will generate 64 blocks by $+1$ modulo 16, where $\infty_j + 1 = \infty_j$ for $j \in \{1, 2, 3, 4\}$.

Example 12. Let $V(K^{(3)}_{5,8,8}) = \mathbb{Z}_{16} \cup \{\infty_1, \infty_2, \infty_3, \infty_4, \infty_5\}$, with vertex partition $\{\{\infty_1, \infty_2, \infty_3, \infty_4, \infty_5\}, \{0, 2, 4, 6, 8, 10, 12, 14\}, \{1, 3, 5, 7, 9, 11, 13, 15\}\}$. Under the action of $\mathbb{Z}_{16}$, all 320 edges of $K^{(3)}_{5,8,8}$ can be divided into 20 orbit representatives. Base blocks are given as follows:

$$\begin{aligned} B = {} & (0, \infty_1, 1, 11, 7, 3), (3, \infty_2, 0, 4, 10, 14), (5, \infty_3, 8, 14, 0, 4), (2, \infty_4, 7, 11, 1, 5), \\ & (4, \infty_5, 3, 1, 9, 13). \end{aligned}$$

Then $T-$ blocks in B cover all the 20 orbit representatives, and will generate 80 blocks by $+1$ modulo 16, where $\infty_j + 1 = \infty_j$ for $j \in \{1, 2, 3, 4, 5\}$.

Example 13. Let $V(K_{12}^{(3)}\backslash K_4^{(3)}) = \mathbb{Z}_8 \cup \{\infty_1, \infty_2, \infty_3, \infty_4\}$, with $\infty_1, \infty_2, \infty_3, \infty_4$ being the vertices of the hole and base blocks are given as follows:

$$B_0 = (0, 5, \infty_1, \infty_3, 1, \infty_2), (0, 2, 7, \infty_1, 5, 6), (1, 2, \infty_1, 5, 3, 7), (4, \infty_2, 3, \infty_1, 6, \infty_3),\\ (\infty_3, 3, 5, \infty_1, 2, \infty_4), (0, \infty_4, 1, 6, 5, \infty_1), (1, 5, \infty_1, \infty_2, \infty_3, \infty_4).$$

$$B_1 = (\infty_2, \infty_4, 0, 5, 4, 6), (\infty_4, \infty_2, 1, 2, 7, 3).$$

Then $T-$ blocks in B_0 cover the 26 orbits, and will generate 52 blocks by $+1$ modulo 8, where $\infty_j + 1 = \infty_j$ for $j \in \{1, 2, 3, 4\}$. The last 2 blocks are listed in B_1.

Example 14. Let $V(K_{13}^{(3)}\backslash K_5^{(3)}) = \mathbb{Z}_8 \cup \{\infty_1, \infty_2, \infty_3, \infty_4, \infty_5\}$, with $\infty_1, \infty_2, \infty_3, \infty_4, \infty_5$ being the vertices in the hole and base blocks are given as follows:

$$B_0 = (0, 1, \infty_1, \infty_3, 6, \infty_2), (0, 3, 6, 7, 1, 4), (0, 2, 6, \infty_1, 1, \infty_2), (0, 3, \infty_2, \infty_1, \infty_4, \infty_3),\\ (3, \infty_1, \infty_3, \infty_4, \infty_2, \infty_5), (0, \infty_3, 2, 4, \infty_2, \infty_4, \infty_5), (2, 6, \infty_3, \infty_4, \infty_1, \infty_2),\\ (\infty_4, 0, 6, 1, \infty_2, \infty_4).$$

$$B_1 = (1, \infty_5, 5, 3, 0, 1), (2, \infty_5, 0, 3, 7, 6), (3, \infty_5, 0, 5, 6, 7), (4, \infty_5, 0, 3, 2, 1), (\infty_5, \infty_2, 1, 2, 5, 3),\\ (\infty_5, 6, 4, 7, 0, 1), (\infty_5, 7, 4, 5, 0, 1), (5, \infty_5, 2, 0, 4, 6), (\infty_2, \infty_5, 6, 0, 7, 4).$$

Then $T-$ blocks in B_0 cover the 30 orbits, and will generate 60 blocks by $+1$ modulo 8, where $\infty_j + 1 = \infty_j$ for $j \in \{1, 2, 3, 4, 5\}$. The last 9 blocks are listed in B_1.

Example 15. Let $V(K_{14}^{(3)}\backslash K_6^{(3)}) = \mathbb{Z}_8 \cup \{\infty_1, \infty_2, \infty_3, \infty_4, \infty_5, \infty_6\}$, with $\infty_1, \infty_2, \infty_3, \infty_4, \infty_5, \infty_6$ being the vertices in the hole and base blocks are given as follows:

$$B_0 = (0, 1, \infty_1, 7, \infty_3, \infty_2), (0, \infty_5, 7, 2, \infty_6, 5), (2, \infty_1, \infty_5, \infty_2, \infty_3, \infty_6), (0, 3, 4, 1, 6, 2),\\ (\infty_4, 0, \infty_5, 6, 1, \infty_3), (0, \infty_6, 1, 3, 6, \infty_4), (4, \infty_2, \infty_5, \infty_3, \infty_6, \infty_4), (1, 5, 6, \infty_5, 2, \infty_6),\\ (1, 3, \infty_3, \infty_1, 7, \infty_2), (0, 5, \infty_2, \infty_3, \infty_1, \infty_4), (0, 4, \infty_1, \infty_3, \infty_2, \infty_4).$$

$$B_1 = (\infty_1, \infty_4, 1, 3, 5, 7), (\infty_1, \infty_4, 2, 6, 8, 4), (\infty_5, \infty_3, 0, 1, 7, 4), (\infty_5, \infty_3, 2, 6, 5, 4),\\ (\infty_6, \infty_3, 1, 5, 0, 7), (\infty_6, \infty_3, 3, 2, 6, 4).$$

Then $T-$ blocks in B_0 cover the 40 orbits, and will generate 80 blocks by $+1$ modulo 8, where $\infty_j + 1 = \infty_j$ for $j \in \{1, 2, 3, 4, 5, 6\}$. The last 6 blocks are listed in B_1.

Example 16. Let $V(K_{15}^{(3)}\backslash K_7^{(3)}) = \mathbb{Z}_8 \cup \{\infty_1, \infty_2, \infty_3, \infty_4, \infty_5, \infty_6, \infty_7\}$, with $\infty_1, \infty_2, \infty_3, \infty_4, \infty_5, \infty_6, \infty_7$ being the vertices in the hole and base blocks are given as follows:

$$B_0 = (0, \infty_2, \infty_5, \infty_3, 6, \infty_7), (0, \infty_5, 7, \infty_3, \infty_1, \infty_7), (0, \infty_4, 1, 5, \infty_2, \infty_5), (0, 1, 6, \infty_2, 7, \infty_7),\\ (2, \infty_1, 3, \infty_4, \infty_2, \infty_6), (0, \infty_7, 5, \infty_4, \infty_3, \infty_1), (1, 5, 2, \infty_2, 6, \infty_5), (0, 3, 4, 7, 1, 5),\\ (1, 3, \infty_7, \infty_1, \infty_4, \infty_5), (2, 6, \infty_1, \infty_4, \infty_7, \infty_3), (4, \infty_6, \infty_4, \infty_2, \infty_3, \infty_5),\\ (0, 5, \infty_1, \infty_2, \infty_5, \infty_3), (0, \infty_3, \infty_1, 2, 7, \infty_4).$$

$B_1 = (1, \infty_5, 3, 5, 0, 1), (2, \infty_5, 0, 3, 7, 6), (3, \infty_5, 0, 5, 6, 7), (4, \infty_5, 0, 3, 2, 1), (5, \infty_5, 2, 0, 4, 6),$
$(\infty_5, 6, 4, 7, 0, 1), (\infty_5, 7, 4, 5, 0, 1), (\infty_6, \infty_7, 1, 2, 5, 3), (\infty_7, \infty_6, 6, 0, 7, 4).$

Then $T-$ blocks in B_0 cover the 48 orbits, and will generate 96 blocks by $+1$ modulo 8, where $\infty_j + 1 = \infty_j$ for $j \in \{1, 2, 3, 4, 5, 6, 7\}$. The last 9 blocks are listed in B_1.

Example 17. Let $V(K_{16}^{(3)} \backslash K_8^{(3)}) = \mathbb{Z}_8 \cup \{\infty_1, \infty_2, \infty_3, \infty_4, \infty_5, \infty_6, \infty_7, \infty_8\}$, with $\infty_1, \infty_2, \infty_3, \infty_4, \infty_5, \infty_6, \infty_7, \infty_8$ being the vertices in the hole and base blocks are given as follows:

$B_0 = (2, \infty_1, \infty_4, \infty_2, \infty_7, \infty_3), (0, 1, \infty_1, \infty_7, \infty_3, \infty_5), (0, 5, 4, 7, 3, 6), (\infty_8, 0, 5, 1, 2, \infty_1),$
$(3, \infty_5, \infty_6, \infty_8, \infty_1, \infty_7), (1, 3, \infty_2, \infty_1, \infty_3, \infty_4), (0, \infty_4, \infty_8, \infty_7, \infty_5, \infty_6),$
$(0, \infty_3, \infty_4, \infty_7, \infty_6, \infty_5), (7, 2, \infty_3, \infty_1, \infty_4, \infty_2), (4, 6, \infty_5, \infty_6, 2, \infty_7),$
$(4, \infty_6, \infty_8, \infty_2, \infty_1, \infty_7), (0, 4, \infty_3, \infty_1, \infty_7, \infty_6), (1, 2, 8, \infty_2, \infty_4, \infty_6),$
$(\infty_2, 0, \infty_7, \infty_3, \infty_5, \infty_4), (1, 5, \infty_2, \infty_5, \infty_4, \infty_8), (0, 3, 4, \infty_5, \infty_7, \infty_6).$

$B_1 = (\infty_2, \infty_8, 1, 3, 5, 7), (\infty_2, \infty_8, 2, 6, 0, 4), (\infty_8, \infty_3, 0, 1, 7, 4), (\infty_8, \infty_3, 2, 6, 5, 4),$
$(\infty_8, \infty_7, 3, 2, 6, 4), (\infty_7, \infty_8, 1, 5, 0, 7).$

Then $T-$ blocks in B_0 cover the 60 orbits, and will generate 120 blocks by $+1$ modulo 8, where $\infty_j + 1 = \infty_j$ for $j \in \{1, 2, 3, 4, 5, 6, 7, 8\}$. The last 6 blocks are listed in B_1.

Example 18. Let $V(K_{11}^{(3)}) = \mathbb{Z}_{10} \cup \{\infty\}$, and let.

$$B_0 = (0, 2, 4, \infty, 9, 6), (1, 4, 5, 9, 6, 8), (1, 2, 5, \infty, 4, 3), (0, 5, 1, 6, 9, 4).$$

$$B_1 = (0, \infty, 3, 5, 6, 4), (1, \infty, 5, 6, 8, 4), (2, \infty, 6, 5, 8, 7), (8, \infty, 2, 3, 5, 4),$$
$$(7, \infty, 1, 4, 0, 3), (9, \infty, 3, 5, 6, 4).$$

Then a maximum $T-$ packing of $K_{11}^{(3)}$ consists of the orbits of the $T-$ blocks in B_0 under the action of the map $\infty \mapsto \infty$, and $j \mapsto j + 1 \pmod{10}$ along with the $T-$ blocks in B_1. Thus, a leave consists of a single edge $\{3, 6, \infty\}$.

Example 19. Let $V(K_{13}^{(3)}) = \mathbb{Z}_{13}$, and let.

$$B_0 = (0, 4, 3, 1, 10, 6), (1, 2, 3, 10, 12, 6), (5, 7, 4, 2, 10, 3), (0, 6, 2, 10, 12, 3).$$

$B_1 = (2, 8, 7, 3, 0, 10), (1, 6, 0, 10, 7, 8), (2, 7, 3, 0, 11, 1, 12), (2, 9, 1, 4, 10, 7), (0, 5, 6, 7, 10, 9),$
$(5, 10, 2, 4, 1, 3), (6, 12, 4, 11, 1, 7), (1, 9, 8, 6, 7, 5), (3, 8, 1, 12, 11, 10), (4, 9, 0, 10, 1, 11),$
$(2, 10, 3, 4, 6, 7), (3, 11, 4, 6, 7, 10), (3, 9, 1, 8, 11, 4), (5, 12, 4, 0, 10, 7), (0, 8, 6, 3, 4, 1),$
$(12, 4, 8, 9, 10, 11), (6, 11, 4, 1, 0, 2), (12, 7, 0, 4, 1, 3), (5, 11, 3, 6, 0, 10).$

Then a maximum $T-$ packing of $K_{11}^{(3)}$ consists of the orbits of the $T-$ blocks in B_0 under the action of the map $j \mapsto j + 1 \pmod{13}$ along with the $T-$ blocks in B_1. Thus, a leave consists of the edge $\{7, 8, 0\}$, $\{5, 8, 0\}$.

Example 20. Let $V(K_{15}^{(3)}) = \mathbb{Z}_{14} \cup \{\infty\}$, and let.

$$B_0 = (0, 1, 3, \infty, 2, 4), (4, 5, 13, 1, 2, 0), (1, 2, 8, 9, 7, 6), (0, 2, 6, 9, 7, 5), (0, 2, 10, 12, 8, \infty), (0, 3, 5, 11, 8, 9), (0, 4, 11, 9, 7, 8).$$

$$B_1 = (\infty, 7, 3, 0, 10, 4), (1, \infty, 4, 10, 12, 8), (\infty, 3, 13, 0, 12, 11), (8, \infty, 3, 5, 11, 4), (5, \infty, 12, 0, 1, 10), (\infty, 11, 0, 7, 5, 6), (\infty, 4, 0, 11, 13, 12), (1, \infty, 6, 7, 9, 11), (\infty, 10, 6, 3, 4, 13), (13, \infty, 7, 8, 9, 5), (2, \infty, 7, 5, 11, 13), (\infty, 2, 6, 10, 8, 7), (\infty, 6, 13, 3, 12, 9), (\infty, 12, 7, 9, 2, 8), (\infty, 9, 5, 3, 4, 0).$$

Then a maximum $T-$ packing of $K_{15}^{(3)}$ consists of the orbits of the $T-$ blocks in B_0 under the action of the map $\infty \mapsto \infty$, and $j \mapsto j + 1 \pmod{14}$ along with the $T-$ blocks in B_1. Thus, a leave consists of the edge $\{0, 6, \infty\}$, $\{8, 0, \infty\}$, $\{10, 0, \infty\}$.

4 Conclusion

Lemma 4.1 There does not exist $S(3, T, 6)$.

Proof. Let $(X, \mathcal{B})$ is an $S(3, T, 6)$, for $v = 6$. For any $x \in X$ denote by $d_i(x)$, $i = 1, 4$, the number of blocks in $\mathcal{B}$ in which the degree of x is i [12]. It follows that $d_1(x) + 4d_4(x) = \frac{(v-1)(v-2)}{2}$. Notice that $d_1(x) + d_4(x) \le |\mathcal{B}|$. When $v = 6$, $|\mathcal{B}| = 5$, solving the above equation, we can get that for any $x \in X$, $d_1(x) = d_4(x) = 2$. It is easy to see that this is impossible. This is a contradiction.

Lemma 4.2 There exists a $T-$ decomposition of $K_v^{(3)}$ if and only if $v = 0, 1, 2, 4, 6, \pmod 8$, $v > 6$.

Proof. According to lemma 1, we can easily know that the necessary conditions for the existence of a $T-$ decomposition of $K_v^{(3)}$ is $v = 0, 1, 2, 4, 6 \pmod 8$, $v \ge 6$. Therefore, we need to prove its sufficiency. Let $v = 8r + l$ where $r \ge 1$ and $l \in \{1, 2, 4, 6, 8\}$. By lemma 2, it suffices to find $T-$ decompositions of $K_{8+l}^{(3)}$, $K_{8+l}^{(3)} \backslash K_l^{(3)}$, $K_{l,8,8}^{(3)} \cup L_{8,8}^{(3)}$ and $K_{8,8,8}^{(3)}$.

We notice that when $l \in \{1, 2\}$, $K_{8+l}^{(3)} \backslash K_l^{(3)}$ is isomorphic to $K_{8+l}^{(3)}$. By example 16, there exists $K_{8+l}^{(3)}$. By example 13, example 15 and example 17, there exists $K_{8+l}^{(3)} \backslash K_l^{(3)}$. Meanwhile, $K_{3,8,8}^{(3)}$ decomposes $K_{6,8,8}^{(3)}$, and $K_{4,8,8}^{(3)}$ decomposes $K_{8,8,8}^{(3)}$. And by example 7–11, there exists $K_{l,8,8}^{(3)} \cup L_{8,8}^{(3)}$, $L_{8,8}^{(3)}$ and $K_{8,8,8}^{(3)}$.

Lemma 4.3 If $v \ge 6$ is an integer, then there exists $MP(3, T, v)$ of $K_v^{(3)}$ where the leave has fewer than four edges.

Proof. According to lemma 4.2, when $v = 0, 1, 2, 4, 6 \pmod 8$, the $T-$ decomposition of $K_v^{(3)}$ is the maximum $T-$ packing which the leave is empty. Hence, we need only consider the maximum $T-$ packing when $v = 3, 5, 7 \pmod 8$. Let $v = 8r + l$ where $r \geq 1$ and $l \in \{3, 5, 7\}$. By Lemma 2, we only need to consider the maximum $T-$ packing of $K_{8+l}^{(3)}$ where a leave has fewer than four edges, and $T-$ decompositions of $K_{8+l}^{(3)} \backslash K_l^{(3)}$, $K_{l,8,8}^{(3)} \cup L_{8,8}^{(3)}$ and $K_{8,8,8}^{(3)}$.

We observe that a $T-$ decomposition of $K_{11}^{(3)} \backslash K_3^{(3)}$ is a subset of a $T-$ packing of $K_{11}^{(3)}$ with a leave consisting of the single edge, then it must be a maximum $T-$ packing of $K_{11}^{(3)}$. By example 14 and example 16, there exists $K_{8+l}^{(3)} \backslash K_l^{(3)}$. $K_{4,8,8}^{(3)}$ decomposes $K_{8,8,8}^{(3)}$, and $K_{7,8,8}^{(3)}$ can be decomposed into copies of $K_{3,8,8}^{(3)}$ and $K_{4,8,8}^{(3)}$. Meanwhile, by example 10, example 11 and example 12, there exists $K_{l,8,8}^{(3)} \cup L_{8,8}^{(3)}$ and $K_{8,8,8}^{(3)}$. And by example 18, example 19, example 20, we can find the maximum $T-$ packing of $K_{11}^{(3)}, K_{13}^{(3)}, K_{15}^{(3)}$.

When $v = 6, 7$, we only need to prove that: $d(3, T, 6) = 3$ and $d(3, T, 7) = 7$. Let $V(K_6^{(3)}) = \mathbb{Z}_6$, then the $P(3, T, 6)$ is as follows: $(0, 1, 2, 3, 4, 5), (2, 4, 0, 1, 3, 5), (3, 5, 2, 4, 0, 1)$. Let $V(K_7^{(3)}) = \mathbb{Z}_7$, then the $P(3, T, 7)$ is as follows: $(1, 3, 0, 2, 5, 4), (0, 2, 4, 5, 6, 3), (2, 5, 1, 3, 4, 6), (1, 6, 2, 4, 5, 0), (2, 6, 2, 5, 0, 1), (4, 6, 2, 5, 0, 3)$

Combining lemma 4.1 and lemma 4.2, we know that there exists an $S(3, T, 6)$ if and only if $v = 0, 1, 2, 4, 6 \pmod 8$, $v > 6$. According to lemma 4.3, for any $v > 6$, there exists $MP(3, T, v)$, and $d(t, T, v) = \left\lfloor \frac{v(v-1)(v-2)}{24} \right\rfloor$. Notice that $d(3, T, 6) = 3$ and $d(3, T, 7) = 7$.

References

1. Zbigniew, L.: Decompositions of hypergraphs into hyper-stars. Discret. Math. **66**, 1–2 (1987)
2. A.F. Mouyart., F. Sterboul.: Decomposition of the complete hypergraph into delta-systems II. J. Comb. Theory Ser. A (41), 139–149 (1986)
3. Z. Lonc.: Solution of a delta-system decomposition problem. J. Comb. Theory, Series A (55), 33–48 (1990)
4. Z. Lonc.: Packing, covering, and decomposing of a complete uniform hypergraph into delta-systems. Graphs and Combinatorics 8, 333–341 (1992)
5. Maximum packing of with stars: Hoffman, D.G. Roberts. Australas. J. Combin. **59**, 206–210 (2014)
6. Bonacini, P., Marino, L.: Edge balanced star-hypergraph designs and vertex colorings of path designs. J. Comb. Des. **30**, 497–514 (2022)
7. Wu, Y., Chang, Y.: Determination of the packing number. Sci. China Ser. A. Math. **52**(11), 2537–2548 (2009)
8. Wu, Y.: Decompositions of 3-uniform hypergraphs and related problems. Beijing JiaoTong University, No. 3 Shangyuan Village (2010). (In chinese)
9. Feng, T., Chai, Z., Chang, Y.: Packings and coverings of complete 3-uniform hypergraph. SCIENTIA SINICA Mathematica **42**(6), 619–633 (2012). (In chinese)
10. Wu, Y., Chang, Y.: On the covering problem. Acta Math. Appl. Sin. **28**, 631–638 (2012)

11. Feng, T., Chang, Y.: Decompositions of 3-uniform hypergraph into hypergraphs of a certain type. Science in China series A **50**, 1035–1044 (2007)
12. Feng, T., Chang, Y.: Decompositions of 3-uniform hypergraph into hypergraph . UTILITAS MATHEMATICA 87, (2012)
13. Armstrong, A., Ryan, C.: On maximum packings of λ -fold complete 3-uniform hypergraphs with triple-hyperstars of size 4. Electronic Journal of Graph Theory and Applications **9**(2), 451–468 (2021)

ANN Based Recognition of the Dimasa Tribal Language of North East India

Maidanon Langthasa, Rupesh Kumar Kanth, and Swanirbhar Majumder

Abstract Speech is a fascinating phenomenon in the human body. Speech differs from one person to another. Most likely, no two people have the same properties of speech like pitch, formant frequencies, duration, etc. Also, different people speak different languages. All these stated properties make speech/language a non-linear and non-stationary signal, which is challenging to implement for a speech/language recognition system. Here we have tried to design the language recognition of the Dimasa language by first implementing a single-word recognition system. This Dimasa language is a Sino-Tibetan language. It is tonal in nature and spoken by a few local tribes in North Eastern part of India. It is predominant in the Districts of Dima Hasao, Karbi Anglong, Nagaon, and Cachar of Assam. Along with the Dimapur District and Jiribam region of Nagaland and Manipur states respectively. Different techniques of Artificial Neural Networks (ANN) were used to obtain different results. The approach is based on feature extraction techniques Mel-Frequency Cepstrum Coefficients (MFCC). In this work, the popular Levenberg–Marquardt training algorithm, Scaled Conjugate Training Algorithm, and Gradient Descent Propagation Training algorithm were used to train for the pattern recognition neural network used to identify the language and speech.

Keywords Speech recognition · Language recognition · ANN · MFCC · Natural language processing · Neural networks

M. Langthasa · R. K. Kanth
Department of ECE, NERIST, Itanagar, India

S. Majumder (✉)
Department of IT, Tripura University, Agartala, India
e-mail: swanirbhar@ieee.org

W. Wang et al. (eds.), *Computational Statistics and Data Intelligence*, Springer Proceedings in Mathematics & Statistics 463,
https://doi.org/10.1007/978-981-97-4438-1_7

1 Introduction

Phonologically, every vowel is a phoneme (although the inverse is not necessarily true), herein a phoneme is simply the smallest unit of sound that carries meaning. The definition of a phoneme is based in the principle of minimal pairs: If there exist two words in a language with a single phonemic alternation, then the two alternating phones (a phone is simply a sound) can be considered to change meaning, and therefore constitute phonemes [1].

Any given vowel possesses a series of acoustic properties called 'formants'; these formants are simply the frequencies (in Hertz) at which the vowel harmonically resonates. These formants can be seen in analyzing a spectrogram, where the formants are simple the darker bars of each vowel. It is also worth noting that the first formant corresponds to the height of a vowel, while the second formant corresponds to the backness of a vowel (the formants modify timbre and can mark phonetic phenomena such asrhoticity, nasalization, and rounding). As shown in Fig. 1, there is a spectrogram analysis of a small speech sample, where the red dots indicate formants (note how they correspond with the darker bars) [2, 2].

For the purposes of this work, a vowel can be defined as follows: a unit of speech that possesses formants and is the nucleus of a syllable. The present study aims at finding out acoustic characteristics of the Dimasa language of Nagaland and Assam for the application of language identification [3, 4].

Here Fig. 2 shows the general process blocks of the system. There are three important modules. First one is pre-processing and automatic segmentation, second one is training and third one is recognition module. Pre-processing and automatic segmentation module works on digitized input speech data and tries to segment input signals into small chunks. These small segments are manually verified by the manual transcribers and mark the correct vowels using Praat software [5]. From the transcribed data vowels' duration and formants (F1 and F2) are extracted and analyzed and Language wise knowledge has been stored for further use. At the time of recognition, input signal is also digitized and passed through pre-processing and segmentation module. These segments are voiced portions containing mainly vowels of the signal and used as the input of language identification block. This block tries to identify the segments of vowels using language knowledge. Decision unit takes the final decision out of all the identified segments and try to converge into final output.

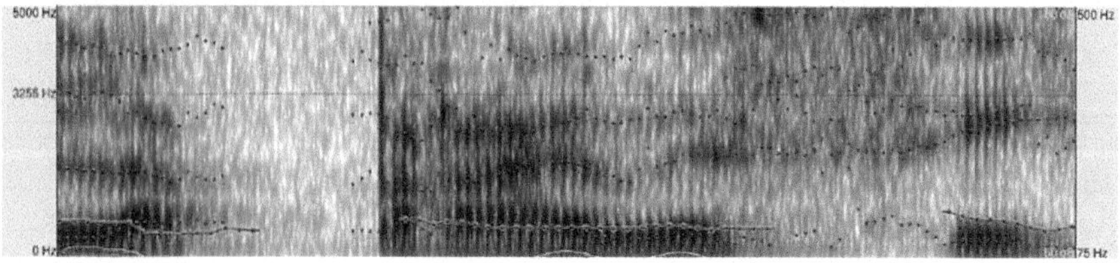

Fig. 1 Spectrogram view of speech signal

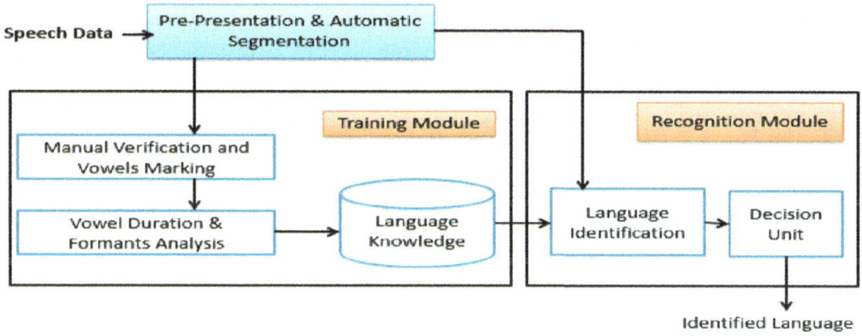

Fig. 2 Block diagram of Language Identification (LID) system

For creation of language knowledge base, text material in target language has to be prepared including digits, paragraph read out, free speech, normal conversations.

2 Literature Review

After going through a review [1] of some language recognition and speech recognition some clarity had been reached. The research on machine based automatic speech recognition has been popular over the decades and ChatGPT has just revolutionized it. This is the amalgamation of technological curiosity with mechanical realization of the various forms of human speech capabilities. The desire for automation via human computer interaction (HCI) for simplification of simple tasks has motivated the researchers to work in this area.

The machine recognition was in existence since 1920 i.e. a century ago. One of the initial attempts was devising an acoustic–phonetic speech recognition system in the 1950s. Davis et al. had built a system in the Bell labs for single speaker recognition of isolated digit. This was based measuring of any digit's spectral resonance at their vowel regions. Olson and Belar at the RCA laboratories, had tried to recognize around 10 distinct syllables for 10 embodied monosyllabic words of a single speaker. Similarly, at the MIT Lincoln laboratories, a vowel recognizer was constructed to recognize 10 vowels embedded by Forgie and Forgie. They did it in a format of /b/ -vowel-/t/ which was independent of speaker.

Several Japanese labs in 1960s and 1970s had started to work in the recognition arena. A hardware-based vowel recognizer was developed in the Radio Research Lab of Tokyo by Suzuki and Nakata. The spectrum analyzer's each channel'soutput nodes were connected to a decision circuit for vowel. In order to choose the spoken vowel a majority decisions logic scheme was used with an elaborate spectrum analyzer filter bank with logic connection. Sakai and Doshita had tried out another hardware effort in Kyoto University in Japan to bild the phoneme recognizing hardware. Here a hardware-based speech segmenter along with the zero-crossing analysis of various spoken data at input regions was used to provide the recognized results for implementation. A standard digit recognizer by Nagata et al. was designed at the Nippon Electric Corporation (NEC) laboratories. Vintsyuk had used methods of dynamic programming. This he proposed for time aligning a pair of speech utterances at Carnegie Mellon University under Reddy's research program. This was later on popular as Dynamic Time Warping (DTW) technique.

Isolated word recognitions were a key research area half a century ago in 1970s [1]. Based on the works of Russians Velichko and Zagoruyko, Japanese Sakoe and Chiba and Americans Itakura and many more researchers. Pattern recognition was used in speech recognition by Velichko and Zagoruyko. The idea of Linear Predictive Coding (LPC) was used for speech recognition systems by usage of appropriate distance measures based on LPC spectral parameters by Itakura. Recognition of connected digits for both speaker dependent and independent systems was proposed by a statistical decision approach by Sambur and Rabiner (1976).

In the 1980s continuous word recognition-based research was under scanner. Rabiner and Wilpon used averaging and clustering techniques for Robust training methods. Here fluently spoken words strings were recognized by concatenated pattern matching in individual words for continuous word recognition system [1]. mixture densities based on hidden Markov models (HMM) were used by Juang, Juang et al. for improving the recognition accuracy large vocabulary speech recognition tasks, independent of speaker at the Bell labs in the middle of 1980. Ma et al. proposed system for combining the good short-time classification virtues of time delay neural networks. These were integrated well with overall recognition capabilities of HMMs for speech recognition (1990–2000s) using hybrid statistical HMM/ANN with connections. Rigoll in 1994 proposed a neural network-based vector quantizer, which was a novel take on hybrid connectionist HMM for speech recognition. Whereas for automatic speech recognition (ASR) Bourlard and Morgan (1998) had used ANN as statistical estimator.

A real time speaker-dependent isolated word recognizer was developed by Pruthi et al. for Hindi. Advancing on this Gupta developed an isolated Hindi word speech recognition system using continuous HMM and analyzes word-based acoustics. For Arabic speech system Al-Qatab and Ainon, in 2010 had discussed development and implementation using HTK for both isolated words and continuous speech system. Very few research works have been done on Dimasa language. The only known works done on Dimasa language are the works of Sarmah [2] along with Caroline Wiltshire who did research on the tone study of Dimasa language and vocal emotion recognition of Dimasa language by Kandali and team [3]. Basu et al. have used similar methods for their work [18].

3 About Dimasa

The North-Eastern part of India happens to be is the eastern-most region of India. Its only connection to Eastern India is via the chicken's neck which is a narrow corridor squeezed between independent nations of Bhutan and Bangladesh. This in the last few years have come up a lot in Indo-China related issues. The North Eastern states of India comprise of the Eight majorly tribal States (Arunachal Pradesh, Assam, Manipur, Meghalaya, Mizoram, Nagaland, Tripura and Sikkim). Northeast India is one of those places where a distinctive number of languages can be found wherein the number of speakers speaking the language is degrading day by day. Due to this, it is important to study the North-East languages and to have a way to treasure them. It is possible to treasure the languages and also to remove the difficulty in understanding a particular language with the help of Digital Signal Processing (DSP) using particularly ANN knowledge.

The different popular tribal languages of Northeast India are: Arunachal Pradesh: Adi, Apatani, Galo, Milang, Nyishi, Nocte, Puimei, Tangsa, Tutsa, Wancho. Assam: Ahom, Amri, Bawm, Biete, Bodo, Bokar, Deori, Dimasa, Kachari, Goalpariya, Karbi, Khelma, Mishing. Manipur: Anal, Chairei, Gangte, Lamkang, Chiru, Chothe, Hmar, Tangkhul, Thadou, Kom, Simte, Maram, Maring, VaipheiPaite, Poumai, and Zou languages. Meghalaya: Amri, Atong, Biete, Khasi, Garo, Jaintia, War, Koch, Pnar. Mizoram: Chakma, Falam, Hakha, Hmar, Lai, Mara, Mizo, Paite, Ralte, Ranglong, Tedim. Nagaland: Angami, Chang, Chiru, Dimasa, Kheza, Khiamnuingam, Konyak, Lamkang, Lotha, Mongsen Ao, Nagamese Creole, Nocte, Ntenyi, Phom, Pochuri, Puiron, Rengma, Sangtam, Sema, Tangkhul, Wancho, Yimchuigam. Sikkim: Nepali, Sherpa, Sikkimese, Gurung, Lepcha, Limbu, Yakkha, Yolmo, Sunwar, Tamang, Magar, Newar. Tripura: Bengali, Kokborok, Bishnupriya Manipuri, Chakma, Darlong, Garo, Bawm, Hrangkhol, Koch, Koloma, Paite, Ralte, Ranglong, Santali.

The Dimasa's which are one of the main stays of Kachari tribes happen to be one of the oldest tribal inhabitants of the North East India. They live mostly in Assam's Dima Hasao District, Karbi Anglong District, Nagaon District, Cachar District and in Dimapur District of Nagaland and Jiribam region and Manipur. According to the census of 2011, there are about 4,20,664 total population of Dimasa. The official language of Dimasa is known as Grao-Dima. The Dimasa language is a tonal Sino-Tibetan language. The Dimasa language has a simple 6 short vowel system as in Table 1. Dimasa has 6 diphthongs but no triphthongs. Here the main vowelsare followed by the glides /i/ and /u/. /ai/ and /au/ are the most common diphthongs. This is tabulated in Table 2. The language does not have triphthongs. Nasal vowels are not present in Dimasa but sometimes it can get nasalized due to regressive assimilation. It has 17 phonemic consonants [4]. The vowel detection is as shown in Fig. 3.

Table 1 Vowels in Dimasa

Unrounded	Front	Central	Back	Rounded
High	I		u	Close
Mid	e	ə	o	Mid-open
Low		a		Open

Table 2 Diphthongs in Dimasa

VG	Dimasa	Gloss
ai	phai	Come
ei	ulei	Similarly
oi	foi	Tolerate
ou	oura	Over there
au	dzau	Stab
ui	hui	Hello

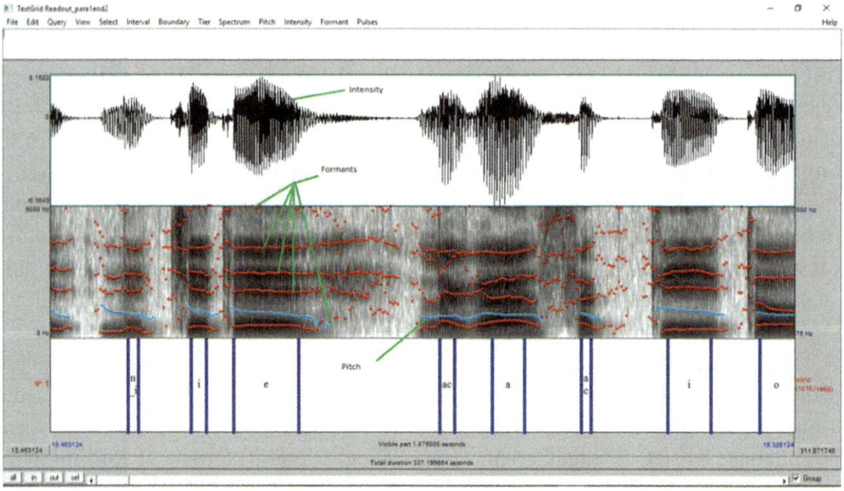

Fig. 3 Labelled diagram of marked vowels

4 Implementation

In this work the Neural network was implemented using MATLAB® Neural Network Toolbox, that provides various methods that can be implemented. The speech/language recognition uses the open-source software Praat to record the audio files. The recording was done in quite condition to reduce the noise. After recording a library called Auditory Toolbox was used that could be linked with MATLAB. This library has the algorithm to calculate the Mel-frequency cepstral coefficient (MFCC) which is used as the unique feature of the recorded audio samples.

4.1 Training Data Implementation

In order to calculate the feature vector, the speech files were recorded and manipulated. The recording was done in 16 kHz sampling frequency and in.wav file format. For removing the unvoiced regions in the speech signals the empty spaces were manually trimmed. This will improve the network error rates.

Using above recording method, we recorded following 10 different Dimasa words which were used to train the network for recognition. These are like: Thang → Go; Jon → Stop; Ao → Yes; Niya → No; Phai → Come; Shisa → Dog; Laisi → Book; Noh → Home; Muthai → Eyes; Yao → Hand; etc.

Ten samples of each word were recorded. Two samples of each word were set aside to test the network. Using the toolbox a function was created to calculate the MFCC of each audio sample. The matrix produced was converted from the above function into a vector which could be fed in neural network.

After extracting the 130 MFCC vectors of each word we created a training matrix. The training matrix was formed of size 130×100. Then we made the target matrix of size 10×100, which are ten classes with 100 words.

4.2 Building the Neural Network

After training matrix and target matrix were ready, we built our pattern recognition neural network. We defined the inputs of network as train matrix and targets as target matrix. We created a variable, hidden layer size and assigned 95 hidden neurons to it. Using the pattern net function, a neural network was created. It was a standard feed forward neural network which would classify inputs according to the provided target values.

Now we divided the input data for testing, validation and testing. 70% of data was used for training, 15%of the data was used for validation, and 15% of the data was used for testing. The different methods of training the network are as under:

- *Conjugate Gradient backpropagation with Powell-Beale (CGB):* The MATLAB network training function is traincgb. It updates the weight and bias values as per conjugate gradient backpropagation with Powell-Beale restarts.
- Conjugate *Gradient backpropagation with Fletcher-Reeves (CGF):* The MATLAB network training function istraincgf.It updates weight and bias values according to the conjugate gradient backpropagation with Fletcher-Reeves updates.

- *Conjugate Gradient Backpropagation with Polak-Ribiere (CGP):* The MATLAB network training function istraincgp. It updates weight and bias values according to the conjugate gradient backpropagation with Polak-Ribiere updates.
- *Levenberg–Marquardt (LM) backpropagation:* The MATLAB network training function istrainlm.It updates weight and bias states according to Levenberg–Marquardt optimization. It is one of the first-choice supervised algorithm and fastest backpropagation algorithmsavailable in the toolbox but requires more memory than other algorithms.
- *Resilient backpropagation (RPROP):* The MATLAB network training function istrainrp. It updates weight and bias values according to the resilient backpropagation algorithm (RPROP).
- *Scaled Conjugate Gradient (SCG) backpropagation:* The MATLAB network training function istrainscg. It updates weight and bias values according to the scaled conjugate gradient method.

Here training occurs as per training parameters provided to the respective functions. Sometimes in absence of any input parameters the default values are used. But these can be overridden once the respective parameter name/value argument pairs are appended to the input argument list, or by appending a structure argument with fields in one or more than one cases.

4.3 Testing the Neural Network

After creating the network and training it the testing function was used that when is called with parameter, testsoundfile and neural network will give the spoken word as output. The flowchart of network testing function is given in Fig. 4.

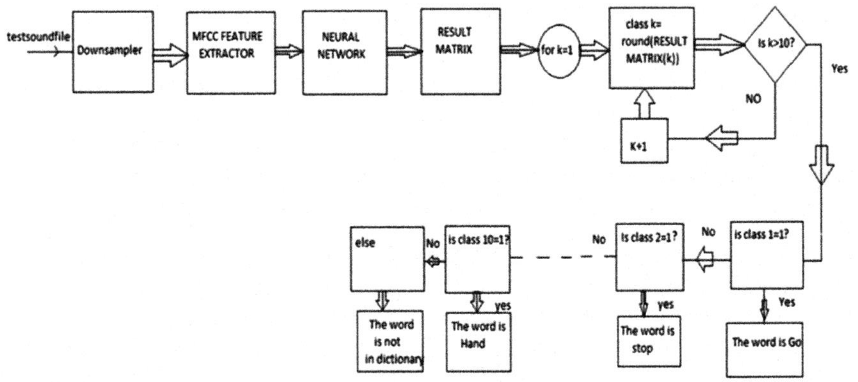

Fig. 4 Flowchart of neural network testing system

5 Results and Discussions

In this work different Training methods were used for training the neural network. The numbers of hidden layers to train the network were also varied. A total of six (06) networks, namely Conjugate Gradient backpropagation with Powell-Beale (CGB), Conjugate Gradient backpropagation with Fletcher-Reeves (CGF), Conjugate Gradient Backpropagation with Polak-Ribiere (CGP), Levenberg–Marquardt (LM) backpropagation, Resilient backpropagation (RPROP) and Scaled Conjugate Gradient (SCG) backpropagation were used. The validation and performance scores of the various ANN training methods employed are given in Table 3. The final best performance figures for the six (06) methods are listed in Figs. 5 and 6.

Comparison between these variations have been with respect to each other were done. Based on this the Table 4 shows the best results employed among all the methods along with overall R value.

Table 3 The validation and performance scores of the various ANN training methods

Method	Single layer neurons			Double layer neurons			Three layer neurons		
	Nos	Validate	Perform	Nos	Validate	Perform	Nos	Validate	Perform
CGB	60	0.80411	0.035447	70,20	0.94941	0.010133	60,30,30	0.9096	0.01788
	70	0.98116	0.004836	90,20	0.97893	0.012581	110,10,10	0.8539	0.04258
	90	0.97618	0.015507	110,20	0.99294	0.000004			
CGF	80	0.92701	0.009279	90,20	0.74326	0.055971	70,20,20	0.7516	0.04289
				100,20	0.90858	0.018675	60,30,30	0.7973	0.0406
CGP	50	0.91997	0.008222	90,20	0.9252	0.022	70,20,30	0.92997	0.01182
	70	0.98985	0.007223	100,20	0.98907	0.006264	60,30,30	0.85029	0.0672
	80	0.91878	0.011221	70,20	0.81323	0.040613			
	90	0.95724	0.014817	50,50	0.94154	0.014281			
	100	0.91598	0.010913						
	120	0.88084	0.020017						
LM	60	0.9647	0.018144	30,40	0.97406	0.023649	20,20,40	0.9401	0.01055
	65	0.99218	0.000171	20,50	0.96287	0.01121	10,20,50	0.90761	0.03726
	70	0.98815	0.005834	40,40	0.94903	0.020203	30,20,50	0.98534	0.00396
	80	0.98976	0.011999	10,60	0.96857	0.011196	30,40,40	0.97452	0.01551
RPROP	60	0.9838	0.011951	40,40	0.9803	0.0065888	20,20,40	0.95528	0.01335
	70	0.97602	0.019138	10,60	0.98311	0.010127	10,20,50	0.99976	0.00019
	80	0.98793	0.003529	20,50	0.97755	0.0041088	30,40,40	0.99327	0.00155
	90	0.98603	0.003699	30,40	0.97405	0.021802	40,40,40	0.97791	0.20648
SCG	85	0.97867	0.014798	30,40	0.96452	0.027067	20,40,40	0.95528	0.02268
	90	0.97264	0.10858	20,60	0.98104	0.0046414	20,50,40	0.99976	0.00415
	110	0.91968	0.0079265	50,50	0.99448	0.0004851	30,50,30	0.99327	0.00457
	115	0.9847	0.0046894	60,30	0.97564	0.015903	40,40,10	0.97791	0.00009

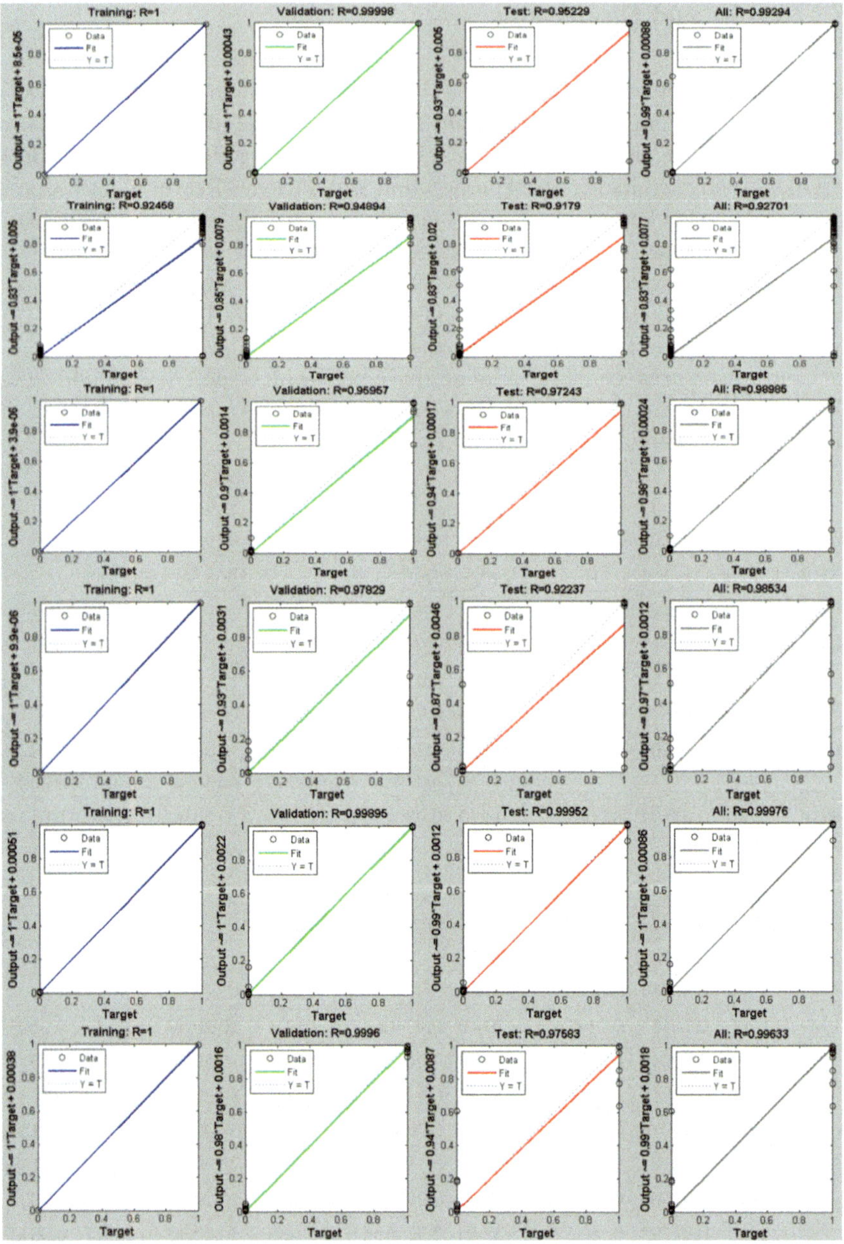

Fig. 5 Training, validation, testing and overall R value for CGB, CGF, CGP, LM, RPROP and SCG respectively

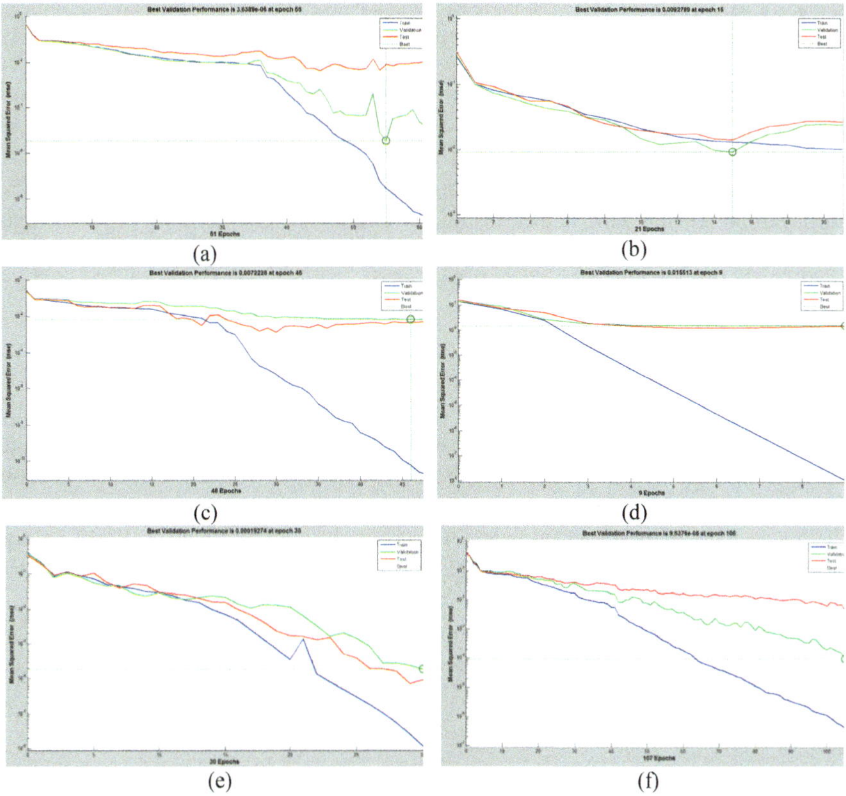

Fig. 6 Performance for **a** Conjugate gradient backpropagation with powell-beale (CGB), **b** Conjugate gradient backpropagation with fletcher-reeves (CGF), **c** Conjugate gradient Backpropagation with polak-ribiere (CGP), **d** Levenberg–marquardt (LM) backpropagation, **e** Resilient backpropagation (RPROP) and **f** Scaled Conjugate Gradient (SCG) backpropagation

Table 4 Table captions should be placed above the tables

Sl	Method	Validation R	All R	Performance
1	CGB	0.99998	0.99294	0.00000364
2	CGF	0.95957	0.98985	0.0072228
3	CGP	0.94894	0.92701	0.0092789
4	LM	0.97829	0.98534	0.0039604
5	RPROP	0.99895	0.99976	0.00019274
6	SCG	0.9996	0.99633	0.0000954

6 Conclusion

In this paper by using the feed forward multilayer perceptron network with Conjugate Gradient back propagation with Powell-Beale (CGB),Conjugate Gradient back propagation with Fletcher-Reeves (CGF), Conjugate Gradient Backpropagation with Polak-Ribiere (CGP), Levenberg–Marquardt (LM) back propagation, Resilient backpropagation (RPROP) and Scaled Conjugate Gradient (SCG) back propagation algorithm were used to recognize the 10 Dimasa words, based on Mel-frequency cepstrum coefficients (MFCC), the sample words were recognized with negligible error. These results show that MFCC can be used to recognize different speech patterns and also show the capabilities of neural networks to classify a pattern. The best result was obtained for the RPROP with overall R being 0.99976.

When trying to do real-time recognition test, network could not recognize all the words correctly. The reason was the presence of unvoiced regions and noise in the recorded speech. The unvoiced region altered the coefficients which caused non-recognition. Speech recognition system remains one of the most important research challenges even after years and years of researches. This is so because there are lots of features in a human speech. Through this work we learned about different neural networks and their training algorithm. This work can be a used as a base for future works on any speech recognition system. Many North-eastern languages of India are yet to be digitalized, so this simple project can help many people to develop their own speech recognition models.

References

1. Ghai, W., Singh, N.: Literature review on automatic speech recognition. Int. J. Comput. Appl. **41**(8), 42–50 (2012)
2. Sarmah, P., Wiltshire, C.: An acoustic study of Dimasa tones. In: Morey, S., Post, M., (eds.), North East Indian Linguistics, pp. 25–44. Foundation Books (2009). https://doi.org/10.1017/UPO9788175968554.004
3. Kandali, A.B., Routray, A., Basu, T.K.: Vocal emotion recognition in five native languages of Assam using new wavelet features. Int. J. Speech Technol. **12**, 1–13 (2009). https://doi.org/10.1007/s10772-009-9046-4
4. Longmailai, M., Cing, Z.N.: Some phonological features of Dimasa and Tedim Chin. North East Indian Linguist. **7**, 15–28 (2015)
5. Praat website 2016.http://www.fon.hun.uva.nl/praat/
6. Davis, S., Mermelstein, P.: Comparison of parametric representations for monosyllabic word recognition in continuously spoken sentences. IEEE Trans. Acoust. Speech Signal Process. **28**, 357–366 (1980). https://doi.org/10.1109/TASSP.1980.1163420
7. Micheva, K.D., Busse, B., Weiler, N.C., O'Rourke, N., Smith, S.J.: Single-synapse analysis of a diverse synapse population: proteomic imaging methods and markers. Neuron **68**(4), 639–653 (2010). https://doi.org/10.1016/j.neuron.2010.09.024
8. Murphy, A.: Implementing Speech Recognition with Artificial Neural Networks (Bachelor's Thesis). Department of Computer Science, Algoma University, Sault Ste, Marie, Ontario (2014)
9. Rashwan, M.S.A., Abo Al Ez, R.H., Abd El Reheem, G., Abd El Reheem, G.: Computational intelligent algorithms for arabic speech recognition. J. Al Azhar Univ. Engin. Sect. **12**(44), 886–893 (2017)

10. Haykin, S.: Neural Networks and Learning Machines [Hardcover]. Pearson Higher Education (2009)
11. Montana, D.J., Davis, L.: Training feedforward neural networks using genetic algorithms. In: Proceedings of the 11th International Joint Conference on Artificial Intelligence - Volume 1 (IJCAI'89), pp. 762–767. Morgan Kaufmann Publishers Inc. (1989)
12. Engelbrecht, A.P.: Computational Intelligence: An Introduction, 2nd edn. Wiley (2007). ISBN: 978-0-470-03561-0
13. Rabiner, L.R., Juang, B.H.: Fundamentals of Speech Recognition. Prentice Hall (1993)
14. Hagan, M.T., Menhaj, M.B.: Training feedforward networks with the Marquardt algorithm. IEEE Trans. Neural Networks **5**(6), 989–993 (1994). https://doi.org/10.1109/72.329697
15. Artificial Neural Networks/Feed-Forward Networks. (n.d.). Wikibooks. https://en.wikibooks.org/wiki/Artificial_Neural_Networks/Feed-Forward_Networks
16. https://engineering.purdue.edu/~malcolm/interval/1998-010/
17. Khan, S., Basu, T., Basu, J., Pal, M., Roy, R., Bepari, M.S.: Data collection and development of Bengali ASR and TTS for conversational AI-based automated advisories in the agriculture domain. In: 2022 4th International Conference on Artificial Intelligence and Speech Technology (AIST), Delhi, India, pp. 1–6 (2022). https://doi.org/10.1109/AIST55798.2022.10065005
18. Basu, J., Basu, T.K., Majumder, S.: Performance evaluation of speaker identification in language and emotion mismatch conditions on Eastern and North Eastern low resource languages of India. In: Nanda, P., Verma, V.K., Srivastava, S., Gupta, R.K., Mazumdar, A.P., (eds.), Data Engineering for Smart Systems, pp. 511–519. Springer. https://doi.org/10.1007/978-981-16-2641-8_49

Missing Value Filling Method Based on Scale Data-Correlation Coefficient Method

Qingcui Dong and Hong Zhao

Abstract Scales are a common form in the preparation of questionnaires. Missing data becomes common in scales for a variety of reasons. The traditional methods of dealing with scale data with missing values usually include direct deletion of cases containing missing values, interpolation using single values, multiple interpolation, etc. In recent years, there has been an increasing number of fill methods for various types of missing data, but few scholars have conducted research on fill methods for missing values specifically for scale data. This paper proposes a method to fill in missing values by using the correlation coefficients between the items of the scale, which is called the correlation coefficient method, in view of the fact that the items of the scale are usually correlated to a certain extent. In this paper, we use some of the scale data obtained from the Q City secondary school students' family education project as an example, and use the mean squared error and mean absolute error as evaluation indicators. By comparison with the common median and plurality methods, it was found that filling the scale data using the correlation coefficient method yielded the smallest mean squared error and mean absolute error, with the median method being the second most effective and the plural method the least effective. Therefore, the correlation coefficient method is more effective and suitable for scale data with large correlations.

Keywords Scale data · Correlation coefficient · Missing value filling

Q. Dong · H. Zhao (✉)
School of Mathematical Sciences, Ocean University of China, Qing Dao, China
e-mail: d17860511351@163.com

W. Wang et al. (eds.), *Computational Statistics and Data Intelligence*, Springer Proceedings in Mathematics & Statistics 463,
https://doi.org/10.1007/978-981-97-4438-1_8

1 Overview of Missing Data

1.1 Missing Data Classification and Processing Methods

There are various types of missing data, which can be classified as follows according to the probability of occurrence of missing data in relation to observed or unobserved variables [1]: Completely random missing [2], Random missing, Non-random missing.

Since most of the research at this stage is on completely random missing data, this paper also reviews and presents its own approach to the common filling methods available for completely random missing data [3].

1.2 Common Methods of Handling Missing Data

For different data types and characteristics, there are various methods to deal with missing values [4]. In recent years, the more common methods for dealing with missing values include case-by-case deletion, single interpolation and multiple interpolation [5]. Different missing value processing methods have their own advantages and disadvantages, and they need to be selected based on different scenarios.

In addition to the missing value methods mentioned above, more accurate and effective methods of filling in missing values can be devised according to real-life situations, taking full account of the characteristics of the data itself. In this paper, for example, we use scale data, and there is usually a high degree of correlation between individual items of the same scale. Therefore, this paper innovatively adopts the correlation coefficient method based on the correlation between data for interpolation [6], with the aim of making full use of the correlation between items and minimising information wastage.

2 Data Sources and Research Methods

2.1 Data Sources

The research data for this paper were obtained from the data of a research project on the home education of secondary school students in Q City. A whole-group sample was taken from eight general secondary schools in four districts of Q City. Two to three classes of the first year of junior secondary school were randomly selected from each school. A total of 932 questionnaires were distributed, and 825 scale questionnaires without missing values were selected as the raw data.

The scale involved in the study was the Core Self-Esteem Scale, which was developed by Judge et al., to measure the level of individuals' core self-esteem, and the Chinese version of the scale was revised by Du Jianzheng and other scholars in China in 2012.

In order to ensure the reliability of the data, the scale was first tested for reliability before interpolation. The metrics is Cronbach coefficient and KMO. The Cronbach coefficient is generally considered to be in the range of 0 to 1, and the closer the value is to 1, the higher the reliability. The Kronbach coefficient of this scale is 0.825, which can be considered as "excellent". In the validity test of the questionnaire, the closer the KMO is to 1, the better, and generally a KMO greater than 0.6 is acceptable. In this test, the KMO of this scale is 0.884, which is greater than 0.6. Therefore, the validity test is also passed. This shows that the scale data used in this paper are true and reliable.

2.2 *Sample Correlation Coefficients and Calculations*

The sample correlation coefficient is used to indicate the degree of linear correlation between variables in a sample, the more common being the Pearson correlation coefficient. It is calculated by the formula:

$$r = \frac{\sum_{i=1}^{n}(x_i - \overline{x})(y_i - \overline{y})}{\sqrt{\sum_{i=1}^{n}(x_i - \overline{x})^2 \cdot \sum_{i=1}^{n}(y_i - \overline{y})^2}} \quad (1)$$

The correlation coefficients of the question items in this dataset ranged from 0.2 to 0.6, all ten variables were positively correlated with each other, and all p-values were less than 0.05, so the correlations between the variables were significant.

The data set selected for this paper consisted of 852 completed questionnaires with 10 question items (variables). The complete data were first programmed in Python to be missing at complete random with specified missing rates, and the correlation coefficients between the variables were calculated and tested for significance. For the complete random missing data set D, if the i-th variable is missing from any of the j-th variables, the case will be deleted to form the complete dataset D_{ij}, and the number of data will be recorded as n_{ij}. For the data in the complete dataset D_{ij}, x_{mi} and x_{mj} denote the value of the i-th variable of the m-th individual and the value of the j-th variable of the m-th individual, respectively. The mean of the i-th variable and the j-th variable are denoted as $\overline{x}_i$ and $\overline{x}_j$, respectively. In this paper, the correlation coefficient between its i-th variable and j-th variable is calculated as:

$$r_{ij} = \frac{\sum_{m=1}^{n_{ij}}(x_{mi} - \overline{x}_i)\left(x_{mj} - \overline{x}_j\right)}{\sqrt{\sum_{m=1}^{n_{ij}}(x_{mi} - \overline{x}_i)}\sqrt{\sum_{m=1}^{n_{ij}}\left(x_{mj} - \overline{x}_j\right)}} \quad (2)$$

2.3 Correlation Coefficient Filling Method and Error Measure for Missing Values

The correlation coefficient is used as the weight in this paper, and the missing values are obtained by weighting all the question items of the same case by a weighted average (if a question item is missing, the weight is 0, if there is no missing, the weight is the corresponding correlation coefficient) and dividing by the sum of the corresponding weights. If the i-th question item of the m-th individual has a missing item, then its predicted value $\hat{y}_{mi}$ can be expressed by the formula

$$\hat{y}_{mi} = \frac{\sum_{j=1,i\neq j}^{10} y_{mj} r_{ij}}{\sum_{j=1,i\neq j}^{10} r_{ij}} \tag{3}$$

y_{mj} is t he value of the jth question item of the mth case

$r_{\_ij}$ is the correlation coefficient between the ith question item and the jth question item.

when y_{mj} is the missing value, then $r_{ij} = 0$ $m = 1, 2, 3 \ldots\ldots n$.

The metrics used to compare the accuracy of missing value filling are: mean squared error (MSE) and mean absolute error (MAE) [7].

In order to reduce the effect of randomness, the random missingness under each missing rate is repeated five times in this paper, and the final filling result is the average of the five filling results. In order to fully compare the effectiveness of different interpolation methods, this paper also uses the median method and the plurality method, which are more common in discrete data, for interpolation under the same missing ratio.

3 Research Process and Results

The complete data was programmed using Python to be missing at different proportions of complete randomness, with the missing proportions being: 3, 5, 10%, 15 and 20%. Each missing proportion was simulated five times, and then filled using the median method, the plurality method and the correlation coefficient method respectively. The MSE and the MAE were used as evaluation indicators to evaluate the processing effects of the three processing methods under different missing proportions, respectively.

The MSE is calculated as:

$$\text{MSE} = \frac{1}{n}\sum_{j=1}^{n} (\hat{y}_j - y_j)^2 \tag{4}$$

The MAE is calculated as:

$$\mathrm{MAE} = \frac{1}{n}\sum_{j=1}^{n} |\hat{y}_j - y_j| \quad (5)$$

where $\hat{y}_j$ denotes the filled value of the jth missing data obtained after filling, y_j denotes the true observed value of the jth missing data in the original complete data set, and n denotes the number of missing data in the whole scale in this simulation. According to the MSE and MAE, the smaller the MSE and MAE, the more accurate the data filling method is and the better the corresponding filling method is.

4 Research Results

Table 1 and Fig. 1 intuitively shows the evaluation indicators of the three filling methods under five different missing rates.

The results show that the MSE and the MAE of the three filling methods, namely the plural method, the median method and the correlation coefficient method, are significantly decreasing, indicating that the filling effect becomes better in turn.

So the plurality method is the least effective, the median method the second most effective and the correlation coefficient method the most effective for the different missing rates.

Table 1 Comparison of evaluation indicators under five different absence rates and three filling methods

Missing rate (%)	Filling methods	MSE	MAE
3	Plural method	2.95	1.22
	Median method	1.89	1.05
	Correlation coefficient method	1.23	0.86
5	Plural method	2.78	1.19
	Median method	1.82	1.02
	Correlation coefficient method	1.21	0.85
10	Plural method	3.05	1.26
	Median method	1.90	1.06
	Correlation coefficient method	1.23	0.86
15	Plural method	2.89	1.20
	Median method	1.96	1.07
	Correlation coefficient method	1.22	0.85
20	Plural method	2.99	1.24
	Median method	1.93	1.06
	Correlation coefficient method	1.29	0.88

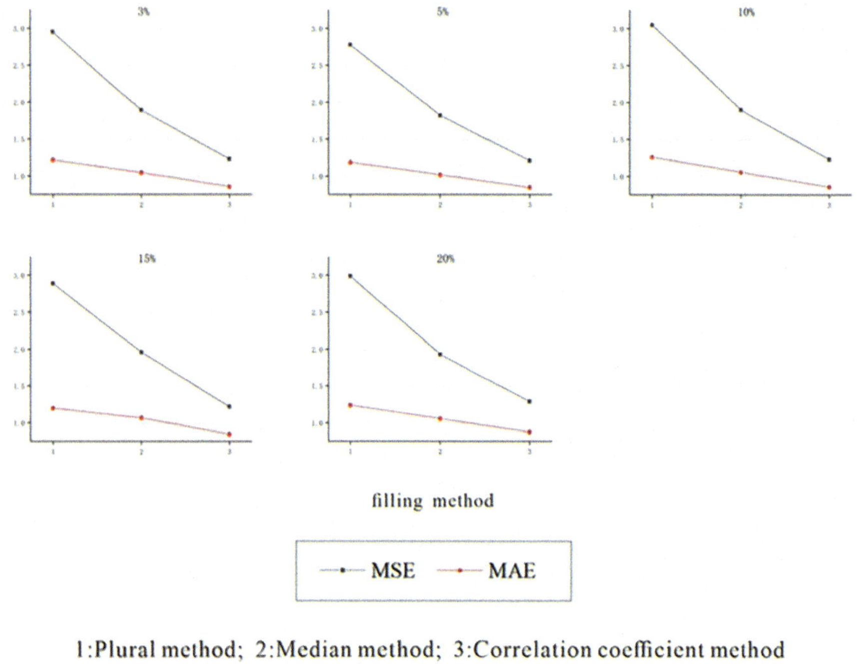

Fig. 1 Comparison of evaluation indicators under three different filling methods

It can be seen that in scale data where correlation between variables is more evident, the use of the correlation coefficient method for filling in missing values is a method that is more accurate.

In this paper, for the scale data, a correlation coefficient method, which uses the correlation coefficients between the individual variables of the scale data to be weighted and divided by the sum of the coefficients to find the missing values, was designed using the correlation coefficients between the individual variables. This method, which makes full use of the information in the scale data, was found to be more accurate in filling in the missing data than the commonly used median and plurality methods by evaluating it with evaluation indicators such as mean squared error and mean absolute error. Therefore, in the process of filling the missing data of a scale, the correlation between the data should be fully considered, and in scales with relatively high correlation between variables, the correlation coefficient method can be chosen to fill the missing values. In the process of filling data in the future, more appropriate methods of filling missing data should also be created according to the characteristics of data from different scales.

Acknowledgements Supported by the Natural Science Foundation of Shandong under project ZR2021MD035, the construction project of teaching case base for professional degree graduate students of Ocean University of China (HDYA22014).

References

1. Fan, Y., Xiao, C.: Several processing techniques for missing data in projects and their effectiveness evaluation: Taking the application of the Chinese version of the Environmental Concern Scale (CNEP) as an example. J. Nanjing Univ. Technol. (Soc. Sci. Ed.). **8**(03), 64–79 (2019). (in chinese)
2. Little, R.J.A.: A test of missing completely at random for multivariate data with missing values. J. Am. Stat. Assoc. **83**(404), 1198–1202 (1998)
3. Jinling Sun.: Research on the processing method of random missing data in psychological questionnaire survey. Zhejiang Norm. Univ. (2017) (in chinese)
4. Leizhen Xia.: Simulation and Comparative Study on Filling in Missing Values of Scales in Questionnaire Surveys. China Medical University. (2018) (in chinese)
5. Pedersen, A.B., Mikkelsen, E.M., Cronin-Fenton, D., et al.: Missing data and multiple imputation in clinical epidemiological research. Clin. Epidemiol. **9**, 157–166 (2017)
6. Wang, Z., Tian, L., Mao, Y.: A discrete data filling algorithm and analysis based on correlation coefficient weighting. Mod. Electron. Technol. **43**(09), 109–112 (2020). (in chinese)
7. Zheng, Z., Chen, Y., Wang, M., Tian, W.: Study on the stability of data filling algorithm under different missing rates. Stat. & Decis. **39**(08), 12–17 (2023). (in chinese)

Sixth-Order Numerical Solver Based on Truncation Error for Solution of Nonlinear Fredholm Equations

Chinedu Nwaigwe, Azubuike Weli, and Dang Ngoc Hoang Thanh

Abstract A sixth order accurate numerical method is proposed for nonlinear integral equations of the Fredholm type. First the problem is discretized and the integral approximated by the classical second order trapezoid rule. Then, the leading truncation error terms of the quadrature rule are approximated. The first derivatives in the error term are approximated using five-point finite difference formula which has fourth order of accuracy, while the third derivative is approximated with a second order finite difference formula. This results to a new leading error term which is of sixth order, hence a sixth order method. Newton algorithm is then designed to approximate the resulting nonlinear system. Numerical examples are used to demonstrate the accuracy and performance of the method. The results are also compared with those from other published works in which the superiority of the proposed method is demonstrated.

Keywords Fredholm integral equations · Finite difference scheme · Trapezoidal rule · Collocation method · Truncation error · High order accuracy · Experimental order of convergence

2020 Mathematics Subject Classifications 65R20 · 45D05

1 Problems

Integral equations are encountered in many applications like in epidemiological modelling [1, 2], antenna wire modelling [3, 4], image processing [5], heat transfer [6], market impact modeling in finance [7], optimal control and economics [8–11] and

C. Nwaigwe (✉) · A. Weli
Department of Mathematics, Rivers State University, Port Harcourt, Nigeria
e-mail: nwaigwe.chinedu@ust.edu.ng

D. N. H. Thanh
Department of Information Technology, School of Business Information Technology, University of Economics Ho Chi Minh City, Ho Chi Minh City, Vietnam
e-mail: thanhdnh@ueh.edu.vn

W. Wang et al. (eds.), *Computational Statistics and Data Intelligence*, Springer Proceedings in Mathematics & Statistics 463,
https://doi.org/10.1007/978-981-97-4438-1_9

physical sciences [12–14]. They also frequently arise from the reformulation of boundary and initial value problems [15, 16].

Due to the numerous applications outlined above, the solution of integral equations are highly desirable. However, in applied settings these equations are usually nonlinear such that closed form analytical solutions are not generally available, hence research in numerical methods for integral equations has gained traction. For instance, many numerical techniques have been proposed through variational iteration methods [17–20], CAS wavelets [21, 22], collocation methods [23, 24], Linear programming [25], Taylor series [26], and Trepezoidal-Newton type method [27–29]. Methods based on fixed point iterations are developed in [16, 30, 31] More ideas can be found in [1, 2, 32–34].

Nwaigwe and Weli [35] considered the following nonlinear Fredholm integral equation:

$$u(x) = p(x) + \int_{y=a}^{b} k(x, y, u(y))\, dy, \qquad x \in [a, b] \subset \mathbb{R}, \tag{1}$$

where $C[a, b] \ni p : \mathbb{R} \to \mathbb{R}, f : \mathbb{R} \times \mathbb{R} \to \mathbb{R}, k : \mathbb{R} \times \mathbb{R} \times \mathbb{R} \to \mathbb{R}$. The work adopted fixed point iteration of Ishikiwa type, and the integral was approximated by using the standard trapezoid rule, leading to a second order method. In [36], the trapezoidal rule was, again, applied to problem (1) in combination with Picard and Newton iterations. This, again, resulted to a second order method. In [37], a nonstandard trapezoid method is proposed and applied to (1). This led to a fourth order method and the associated nonlinear system was approximated by using four different iterative schemes. The results clearly demonstrated the fourth order accuracy of the proposed method.

To the best of our knowledge, there is currently no sixth order numerical method, for problem (1), that is based on the truncation errors of the classical second order trapezoidal rule. This is the gap the present study fills. Consequently, in Sect. 2, we first recall the trapezoid rule with its truncation error up to the fourth order term. Then, we derive the sixth order method in detail. Numerical examples are presented in Sect. 3 to demonstrate the high order accuracy of the method and its high superiority over the method of [35]. We make concluding remarks in Sect. 4.

2 The Numerical Method

The numerical algorithm to solve problem (1) is presented in this section. Let $N \in \mathbb{Z}^+, N > 4$ be an integer, $h := \frac{b-a}{N}$, and the mesh be $\Omega_h = \{x_i = a + ih : i = 0, 1, \cdots, N\}$. Define the grid functions

$$u_i \approx u(x_i),\ \xi_i^N := h \begin{cases} \frac{1}{2}, \text{ if } i = 0, N, \\ 1, \text{ if } 0 < i < N, \\ 0, \text{ otherwise}, \end{cases} \quad \text{for each } i. \tag{2}$$

The following Lemmas will be needed in the numerical formulation below.

Lemma 2.1 (Trapezoid Rule, [38]) *Let* $f \in C^5(a,b)$, *and* h, N *and* $x_j \in \Omega_h$, *for* $j = 0, 1, \cdots, N$, *be defined as above, then*

$$\int_a^b f(x)\,dx = \sum_{k=0}^{N} \xi_k^N f(x_k) + \frac{h^2}{12}\Big[f'(x_0) - f'(x_N)\Big] - \frac{h^4}{720}\Big[f^{(3)}(x_0) - f^{(3)}(x_N)\Big] + O(h^6). \tag{3}$$

Lemma 2.2 (See [39, 40]) *Let* $0 < h \in \mathbb{R}$,

(i) if $f \in C^5(x_i - 4h, x_i)$, *then*

$$f'(x_i) = \frac{1}{12h}\Big(25f(x_i) - 48f(x_i - h) + 36f(x_i - 2h) (x_i - 3h) + 3f(x_i - 4h)\Big) + O(h^4), \tag{4}$$

(ii) if $f \in C^5(x_i, x_i + 4h)$, *then*

$$f'(x_i) = \frac{1}{12h}\Big(-25f(x_i) + 48f(x_i + h) - 36f(x_i + 2h) + 16kf(x_i + 3h) - 3f(x_i + 4h)\Big) + O(h^4), \tag{5}$$

(iii) Backward Difference Scheme:

$$f^{(3)}(x_i) = \frac{1}{2h^3}\Big(5f(x_i) - 18f(x_i - h) + 24f(x_i - 2h) - 14kf(x_i - 3h) + 3f(x_i - 4h)\Big) + O(h^2), \tag{6}$$

and

(iv) Forward Difference Scheme:

$$f^{(3)}(x_i) = \frac{1}{2h^3}\Big(-5f(x_i) + 18f(x_i + h) - 24f(x_i + 2h) + 14kf(x_i + 3h) - 3f(x_i + 4h)\Big) + O(h^2). \tag{7}$$

2.1 The Algorithm

Collocating (1) at $x_i \in \Omega$, we get

$$u(x_i) - p(x_i) - \int_a^b k(x_i, y, u(y))dy = 0, \text{ for } x_i \in \Omega_h. \tag{8}$$

Using (3)–(7), then (8) is approximated as

$$u_i - p(x_i) - \bar{I}_i^h = 0, \quad \forall i = 0, 1, \cdots, N, \tag{9}$$

where $\bar{I}_i^h$ is a sixth order approximation of the integral $\int_a^b k(x, y, u(y))\,dy$ and is define as

$$\begin{aligned}
\bar{I}_i^h =& \sum_{j=0}^{N} \xi_i^N k(x_i, x_j, u_j) \\
&+ \frac{h}{144}\Bigg[\Bigg(- 25k(x_i, x_0, u_0) + 48k(x_i, x_1, u_1) \\
&- 36k(x_i, x_2, u_2) + 16k(x_i, x_3, u_3) - 3k(x_i, x_4, u_4)\Bigg) \\
&- \Bigg(25k(x_i, x_N, u_N) - 48k(x_i, x_{N-1}, u_{N-1}) \\
&+ 36k(x_i, x_{N-2}, u_{N-2}) - 16k(x_i, x_{N-3}, u_{N-3}) \\
&+ 3k(x_i, x_{N-4}, u_{N-4})\Bigg)\Bigg] \\
&- \frac{h}{1440}\Bigg[\Bigg(- 5k(x_i, x_0, u_0) + 18k(x_i, x_1, u_1) \\
&- 24k(x_i, x_2, u_2) + 14k(x_i, x_3, u_3) - 3k(x_i, x_4, u_4)\Bigg) \\
&- \Bigg(5k(x_i, x_N, u_N) - 18k(x_i, x_{N-1}, u_{N-1}) \\
&+ 24k(x_i, x_{N-2}, u_{N-2}) - 14k(x_i, x_{N-3}, u_{N-3}) \\
&+ 3k(x_i, x_{N-4}, u_{N-4})\Bigg)\Bigg].
\end{aligned} \tag{10}$$

By using (3)–(7), one can show that

$$\int_a^b k(x, y, u(y))\,dy = \bar{I}_i^h + O(h^6), \tag{11}$$

hence the algorithm (9)–(10) is a sixth order approximation of (1).

The scheme (9)–(10) is an $(N+1)$-by-$(N+1)$ system for $(N+1)$ nonlinear algebraic equations, so it requires some form of iterative scheme to approximate the solution. Our experience and results in [36, 37] show that the contraction requirement of Banach fixed point theorem may not be satisfied by this approximation, this is especially due to the terms outside the summation sign in $\bar{I}^h, i$. Hence, we approach the approximation of (9)–(10) via Newton method. We use the scheme

$$\mathbf{u}^{m+1} = \mathbf{u}^m - J^{-1}(\mathbf{u}^m)\mathbf{F}(\mathbf{u}^m), \quad m \geq 0, \tag{12}$$

where $\mathbf{u}^m = (u_0, u_1, \cdots, u_N)^T$ at the m-th iteration, $\mathbf{F}(\mathbf{u}^m)$ is a vector with i-th components $f_i(\mathbf{u}^m) = u_i - p_i - \bar{I}_i^h$, $i = 0, 1, \cdots, N$, and J is the Jacobian Matrix of $\mathbf{F}$.

Finally, the scheme is initialized with the initial guess:

$$u_i = p_i + \bar{H}_i^h, \quad i = 0, 1, 2, \cdots, N, \tag{13}$$

where $\bar{H}_i^h$ is given as

$$\begin{aligned}
\bar{H}_i^h = &\sum_{j=0}^{N} \xi_i^N k(x_i, x_j, p_j) \\
&+ \frac{h}{144}\Bigg[\Big(-25k(x_i, x_0, p_0) + 48k(x_i, x_1, p_1) \\
&- 36k(x_i, x_2, p_2) + 16k(x_i, x_3, p_3) - 3k(x_i, x_4, p_4)\Big) \\
&- \Big(25k(x_i, x_N, p_N) - 48k(x_i, x_{N-1}, p_{N-1}) \\
&+ 36k(x_i, x_{N-2}, p_{N-2}) - 16k(x_i, x_{N-3}, p_{N-3}) \\
&+ 3k(x_i, x_{N-4}, p_{N-4})\Big)\Bigg] \\
&- \frac{h}{1440}\Bigg[\Big(-5k(x_i, x_0, p_0) + 18k(x_i, x_1, p_1) \\
&- 24k(x_i, x_2, p_2) + 14k(x_i, x_3, p_3) - 3k(x_i, x_4, p_4)\Big) \\
&- \Big(5k(x_i, x_N, p_p) - 18k(x_i, x_{N-1}, p_{N-1}) \\
&+ 24k(x_i, x_{N-2}, p_{N-2}) - 14k(x_i, x_{N-3}, p_{N-3}) \\
&+ 3k(x_i, x_{N-4}, p_{N-4})\Big)\Bigg].
\end{aligned} \tag{14}$$

This completes the numerical formulation; numerical examples are presented in the following section to demonstrate the accuracy and convergence of the proposed numerical scheme in solving problem (1).

3 Numerical Experiments

Six examples are provided in this section to demonstrate the accuracy of the proposed method. All the problems in this section are designed in [35], see also [36, 37]. They derived via manufactured solutions technique [41–47]. The solution computed using the proposed sixth order method is compared with the second order method proposed in [35]. The results are shown in Tables 1, 2, 3, 4, 5 and 6. To minimize typing, in Tables 1, 2, 3, 4, 5 and 6 we use the symbols **Error-New Scheme** and **Error NW-M** to mean the errors computed with the newly proposed method and the error computed by the method of Nwaigwe and Weli in [35]. Similarly, the symbols **EOC-New Scheme** and **EOC NW-M** are the experimental order of convergence (EOC) observed for the new method and that of [35] respectively.

Example 1 The first example is, [35]:

$$u(x) = \frac{\frac{-x}{6} + (x^2+1)cos^{-1}(x^2)}{1+x^2} + \int_0^1 \frac{xy^3 cos(u)}{1+x^2} dy,$$

whose exact solution is $u(x) = cos^{-1}(x^2)$. This problem is solved using our proposed method and that of [35] on grids with 20, $\cdots$, 640 grid points. The results are shown in Table 1. One can see that the new method computes the exact solution to sixteen decimal places even with a very coarse grid with only 20 points. Also, one can see

Table 1 Results for Problem 1, N = number of sub-intervals

N	Error-new scheme	Error NW-M	EOC-new scheme	EOC NW-M
20	4.44089e-16	5.36470e-04	–	–
40	4.44089e-16	1.27194e-04	0.0	2.08
80	4.44089e-16	3.09859e-05	0.0	2.04
160	4.44089e-16	7.64818e-06	0.0	2.02
320	4.44089e-16	1.89997e-06	0.0	2.01
640	4.44089e-16	4.73497e-07	0.0	2.0

Table 2 Results for Problem 2, N = number of sub-intervals

N	Error-new scheme	Error NW-M	EOC-new scheme	EOC NW-M
5	6.66933e-04	2.15087e-03	–	–
10	1.76681e-04	3.62216e-04	1.92	2.57
20	1.82645e-06	7.94878e-05	6.6	2.19
40	1.58763e-08	1.87797e-05	6.85	2.08
80	5.12250e-10	4.57195e-06	4.95	2.04
160	9.23894e-12	1.12836e-06	5.79	2.02
320	1.50657e-13	2.80307e-07	5.94	2.01
640	2.44249e-15	6.98566e-08	5.95	2.0

Table 3 Results for Problem 3, N = number of sub-intervals

N	Error-new scheme	Error NW-M	EOC-new scheme	EOC NW-M
5	1.59860e-06	2.16760e-04	–	–
10	4.63788e-08	4.28430e-05	5.11	2.34
20	6.04955e-10	9.61399e-06	6.26	2.16
40	8.26172e-12	2.28187e-06	6.19	2.07
80	1.19738e-13	5.56120e-07	6.11	2.04
160	1.88738e-15	1.37287e-07	5.99	2.02
320	2.22045e-16	3.41070e-08	3.09	2.01
640	2.22045e-16	8.50013e-09	0.0	2.0

Table 4 Results for Problem 4, N = number of sub-intervals

N	Error-new scheme	Error NW-M	EOC-new scheme	EOC NW-M
5	1.30837e-04	3.39985e-03	–	–
10	5.12320e-06	6.79933e-04	4.67	2.32
20	8.70988e-08	1.52896e-04	5.88	2.15
40	1.34200e-09	3.63064e-05	6.02	2.07
80	2.05060e-11	8.84925e-06	6.03	2.04
160	3.16025e-13	2.18463e-06	6.02	2.02
320	5.44009e-15	5.42741e-07	5.86	2.01
640	1.33227e-15	1.35260e-07	2.03	2.0

Table 5 Results for Problem 5, N = number of sub-intervals

N	Error-new scheme	Error NW-M	EOC-new scheme	EOC NW-M
5	5.27024e-04	4.89373e-03	–	–
10	3.66384e-05	9.66913e-04	3.85	2.34
20	5.43310e-07	2.16895e-04	6.08	2.16
40	3.75463e-09	5.14755e-05	7.18	2.08
80	2.28562e-11	1.25450e-05	7.36	2.04
160	1.15019e-13	3.09690e-06	7.63	2.02
320	2.22045e-16	7.69377e-07	9.02	2.01
640	2.22045e-16	1.91742e-07	0.0	2.0

that the solution computed by the new method on a 20-point grid is far better (more accurate) than the solution computed by NW-M on a very fine grid of 640 points. The numerical solution converged to the exact solution very fast—on a very few grid refinements, hence the EOC became zero since the errors is constant (solution converged). That is the reason why the EOCs are all zero for the new method in Table 1. Figure 1 also shows the plots of the solution computed by the new method on 10 and 20 points grids. The high accuracy of the proposed method is evidenced.

Table 6 Results for Problem 6, N = number of sub-intervals

N	Error-new scheme	Error NW-M	EOC-new scheme	EOC NW-M
5	1.45507e-04	2.43664e-01	–	–
10	4.59597e-06	3.42621e-02	4.98	2.83
20	6.67170e-08	7.09878e-03	6.11	2.27
40	9.75447e-10	1.65782e-03	6.1	2.1
80	1.46723e-11	4.02516e-04	6.05	2.04
160	2.26152e-13	9.92719e-05	6.02	2.02
320	8.10463e-15	2.46535e-05	4.8	2.01
640	3.88578e-15	6.14124e-06	1.06	2.01

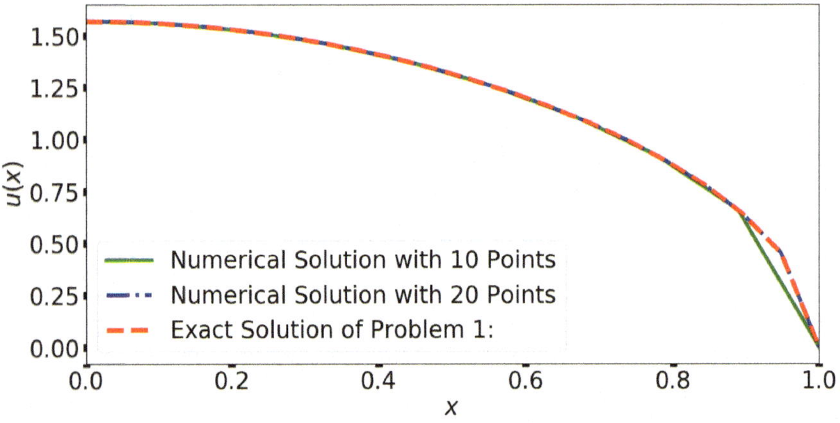

Fig. 1 Plot of exact solution and numerical solution for problem 1

Example 2 The second example is, [35]:

$$u(x) = x^2\left(x - \frac{\pi\sqrt{3}}{54}\right) + \int_0^1 \frac{x^2y^2u}{1+u^2+u^4}dy$$

with exact solution is $u(x) = x^3$.

The results of solving this problem using the proposed method and NW-M are displayed in Table 2. One can see that the sixth order accuracy of the new method is evidenced, and the solution computed by the proposed method on a grid of 40 points is more accurate than that computed by NW-M on a grid of 640 points. The plots of the solution computed by the new method and the exact solution are shown in Fig. 2.

Example 3 The next example is

$$u(x) = -\frac{x^2}{12} + sin(x) - \frac{sin(2)}{48} - \frac{1}{54} - \frac{cos(3)}{432}$$

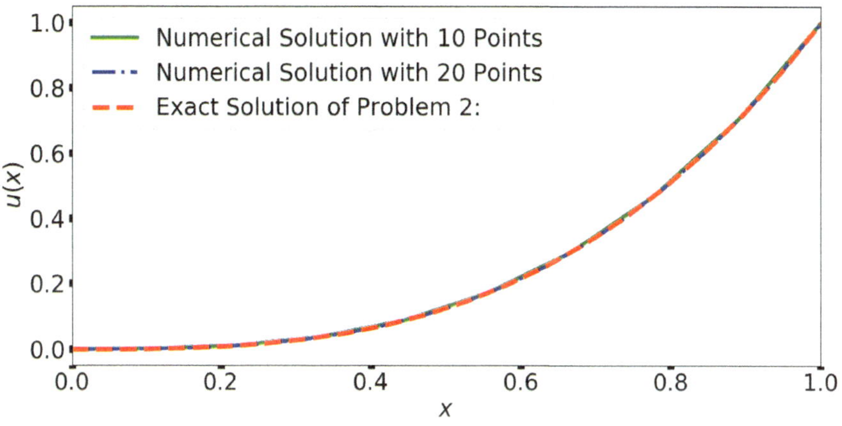

Fig. 2 Plot of exact solution and numerical solution for problem 2

$$+\frac{cos(1)}{48}+\frac{1}{12}\int_{0}^{1}\left(x^2+y+\frac{u^3}{3}-u^2\right)dy,$$

with exact solution $u(x) = sin(x)$. Table 3 shows the results computed for this example using the new method and NW-M on different grids. One can see that the solution computed by the new method on a coarse grid of 20 points is more accurate than that computed by NW-M on a fine grid of 640 point. Figure 3 shows the plots of the solution computed by the new method. The high order accuracy of the method is clearly demonstrated.

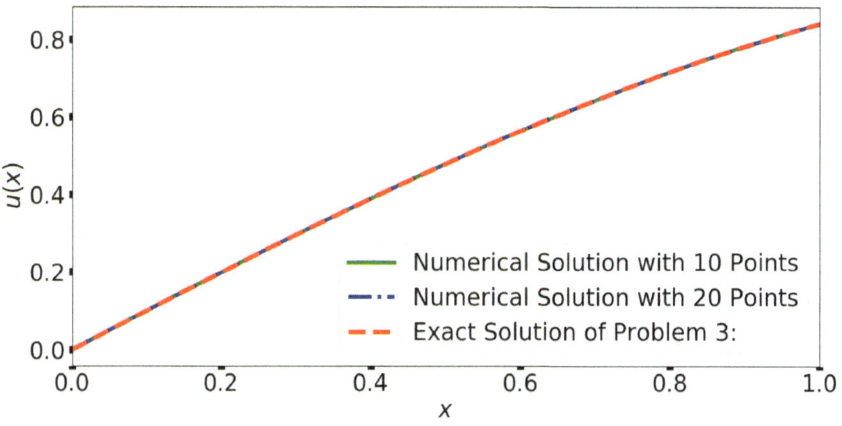

Fig. 3 Plot of exact solution and numerical solution for problem 3

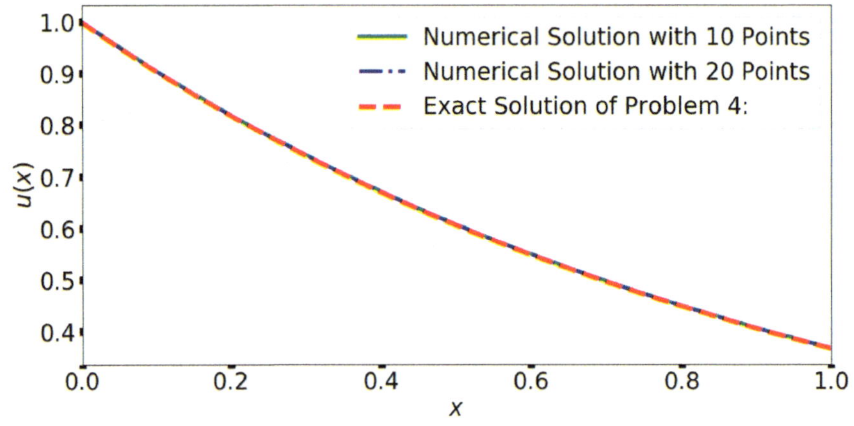

Fig. 4 Plot of exact solution and numerical solution for problem 4

Example 4 Next example is

$$u(x) = \frac{e^{x-1}}{2} - \frac{1}{2}e^{x-e^2} + e^{-x} + \int_0^1 ue^{x-y-u^2}dy$$

with exact solution $u(x) = e^{-x}$, see [35]. The results computed for this example are shown in Table 4 and Fig. 4. Table 4 shows that the new method computes very high accurate solution. In particular, the solution computed by the new method on 20 grid points is better than that of NW-M on 640 grid point.

Example 5 The fifth example is, [35]:

$$u(x) = \frac{x^3(6 - log(2))}{6} + \int_0^1 \frac{x^3 y^5}{1 + u^2(y)}dy.$$

with exact solution $u(x) = x^3$. It be seen in Fig. 5 that the sixth order accuracy of the new method is demonstrated and its superiority over NW-M is obvious.

Example 6 The last example is

$$u(x) = -x^4 + \cos(x) - 5/6 - \frac{\sin(2)}{4} + \frac{\sin(3)}{36} + \frac{\sin(1)}{4} + \int_0^1 (x^4 + y^2 - u^3/3 + u^2)dy,$$

with exact solution is $u(x) = \cos(x)$, see [35]. Table 6 shows the high order accuracy of the new method and that the solution computed by the new method on a coarse grid with only 10 points is better than that computed with NW-M on a mesh of 640 grid points.

4 Conclusion

This paper suggests a new numerical method for solving nonlinear Fredholm integral equations. The major goal was to derive a method which has high order of accuracy. Hence, a sixth-order accurate numerical algorithm is successfully constructed, implemented and tested for nonlinear Fredholm equations. The method is based on simple classical trapezoid rule and one-sided finite difference approximations. The method is applied to six example problems and the results are compared with those of an existing method proposed in [35]. The results show that the proposed method

(i) is highly accurate,
(ii) is even more accurate on a coarse grid than the compared existing method [35] on a finer grid, and
(iii) the method is efficient.

It is therefore concluded that a careful treatment of the truncation error of classical quadrature rules can lead to very robust and accurate solution of integral equations.

Acknowledgements Dr. Dang Ngoc Hoang Thanh appreciates the University of Economics Ho Chi Minh City (UEH) for supporting this research.

Conflict of Interest
There is no conflict of interest.
Funding
Not Applicable.
Informed Consent Statement
Not Applicable.
Data Availability Statement
Not Applicable.

References

1. Maleknejad, K., Hadizadeh, M.: A new computational method for volterra-fredholm integral equations. Comput. Math. Appl. **37**(9), 1–8 (1999)
2. Wazwaz, A.-M.: A reliable treatment for mixed volterra-fredholm integral equations. Appl. Math. Comput. **127**(2–3), 405–414 (2002)
3. Alipanah, A.: Solution of hallen's integral equation by using radial basis functions. Math. Rep. **15**(3), 211–220 (2013)
4. Mei, K.: On the integral equations of thin wire antennas. IEEE Trans. Antennas Propag. **13**(3), 374–378 (1965)
5. Yao, L., Shen, L., Yuesheng, X.: Integral equation models for image restoration: high accuracy methods and fast algorithms. Inverse Prob. **26**(4), 045006 (2010)
6. Olmstead, W.E., Handelsman, R.A.: Asymptotic solution to a class of nonlinear volterra integral equations. ii. SIAM J. Appl. Math. **30**(1), 180–189 (1976)
7. Gatheral, J., Schied, A., Slynko, A.: Transient linear price impact and fredholm integral equations. Math. Financ.: Int. J. Math., Stat. Financ. Econ. **22**(3), 445–474 (2012)
8. Hu, S., Khavanin, M., Zhuang, W.A.N.: Integral equations arising in the kinetic theory of gases. Appl. Anal. **34**(3–4), 261–266 (1989)

9. Jerri, A.: Introduction to Integral Equations with Applications. Wiley (1999)
10. Oregan, D.: Existence results for nonlinear integral equations. J. Math. Anal. Appl. **192**(3), 705–726 (1995)
11. Prüss, J.: Evolutionary Integral Equations and Applications, vol. 87. Birkhäuser (2013)
12. Abdou, M.A.: On a symptotic methods for fredholm-volterra integral equation of the second kind in contact problems. J. Comput. Appl. Math. **154**(2), 431–446 (2003)
13. Le, T.-D., Moyne, C., Murad, M.A., Lima, S.A.: A two-scale non-local model of swelling porous media incorporating ion size correlation effects. J. Mech. Phys. Solids **61**(12), 2493–2521 (2013)
14. Rocha, A.C., Murad, M.A., Moyne, C., Oliveira, S.P., Le, T.D.: A new methodology for computing ionic profiles and disjoining pressure in swelling porous media. Comput. Geosci. **20**(5), 975–996 (2016)
15. Wazwaz, A.-M.: First Course in Integral Equations, A. World Scientific Publishing Company (2015)
16. Micula, S.: On some iterative numerical methods for mixed volterra-fredholm integral equations. Symmetry **11**(10), 1200 (2019)
17. Lan, X.: Variational iteration method for solving integral equations. Comput. Math. Appl. **54**(7–8), 1071–1078 (2007)
18. Sheth, S.S., Singh, D.T., et al.: An analytical approximate solution of linear, system of linear and non linear volterra integral equations using variational iteration method. In: Proceedings of International Conference on Advancements in Computing and Management (ICACM) (2019)
19. Yousefi, S.A., Lotfi, A., Dehghan, M.: He's variational iteration method for solving nonlinear mixed volterra-fredholm integral equations. Comput. Math. Appl. **58**(11–12), 2172–2176 (2009)
20. Hamoud, A., Ghadle, K.: On the numerical solution of nonlinear volterra-fredholm integral equations by variational iteration method. Int. J. Adv. Sci. Tech. Res. **3**, 45–51 (2016)
21. Ezzati, R., Najafalizadeh, S.: Numerical methods for solving linear and nonlinear volterra-fredholm integral equations by using cas wavelets. World Appl. Sci. J. **18**(12), 1847–1854 (2012)
22. Shiralashetti, S.C., Lamani, L.: Cas wavelets stochastic operational matrix of integration and its application for solving stochastic itô-volterra integral equations. Jordan J. Math. Stat. (JJMS) **14**(3), 555–580 (2021)
23. Ordokhani, Y., Razzaghi, M.: Solution of nonlinear volterra-fredholm-hammerstein integral equations via a collocation method and rationalized haar functions. Appl. Math. Lett. **21**(1), 4–9 (2008)
24. Brunner, H.: On the numerical solution of nonlinear volterra-fredholm integral equations by collocation methods. SIAM J. Numer. Anal. **27**(4), 987–1000 (1990)
25. Hasan, P.M.A., Suleiman, N.A.: Numerical solution of mixed volterra-fredholm integral equations using linear programming problem. Appl. Math. **8**(3), 42–45 (2018)
26. Chen, Z., Jiang, W.: An approximate solution for a mixed linear volterra-fredholm integral equation. Appl. Math. Lett. **25**(8), 1131–1134 (2012)
27. Nwaigwe, C., Benedict, D.N.: Generalized banach fixed-point theorem and numerical discretization for nonlinear volterra-fredholm equations. J. Comput. Appl. Math. **425**, 115019 (2023)
28. Bazm, S., Lima, P., Nemati, S.: Analysis of the euler and trapezoidal discretization methods for the numerical solution of nonlinear functional volterra integral equations of urysohn type. J. Comput. Appl. Math. 113628 (2021)
29. Nwaigwe, C.: Solvability and approximation of nonlinear functional mixed volterra-fredholm equation in banach space. J. Integral Equ. Appl. **34**(4), 489–500 (2022)
30. Nwaigwe, C., Micula, S.: Fast and accurate numerical algorithm with performance assessment for nonlinear functional volterra equations. https://www.researchgate.net/publication/361722935_Fast_and_Accurate_Numerical_Algorithm_with_Performance_Assessment_for_Nonlinear_Functional_Volterra_Equations, 2023. (Accessed on 16-February-2023)

31. Micula, S.: Numerical solution of two-dimensional fredholm-volterra integral equations of the second kind. Symmetry **13**(8), 1326 (2021)
32. Aziz, I., et al.: New algorithms for the numerical solution of nonlinear fredholm and volterra integral equations using haar wavelets. J. Comput. Appl. Math. **239**, 333–345 (2013)
33. Atkinson, K.E.: The numerical solution of integral equations of the second kind. In: Cambridge Monographs on Applied and Computational Mathematics (1996)
34. Youssri, Y.H., Hafez, R.M.: Chebyshev collocation treatment of volterra-fredholm integral equation with error analysis. Arabian J. Math. **9**(2), 471–480 (2020)
35. Nwaigwe, C., Weli, A.: Ishikawa-collocation method for nonlinear fredholm equations with non-separable kernels. J. Adv. Math. Comput. Sci. **38**(3), 1–11 (2023)
36. Weli, A., Nwaigwe, C.: Computational analysis of two numerical solvers for functional fredholm equations. Int. J. Math. Comput. Methods **8**, 1–8 (2023)
37. Nwaigwe, C., Weli, A., Thanh, D.N.H.: Fourth-order trapezoid algorithm with four iterative schemes for nonlinear integral equations (2023). https://www.researchgate.net/profile/Chinedu-Nwaigwe/publication/368830972_Fourth-Order_Trapezoid_Algorithm_with_Four_Iterative_Schemes_for_Nonlinear_Integral_Equations/links/63fc62685749505945489d96/Fourth-Order-Trapezoid-Algorithm-with-Four-Iterative-Schemes-for-Nonlinear-Integral-Equations.pdf
38. Fornberg, B.: Improving the accuracy of the trapezoidal rule. SIAM Rev. **63**(1), 167–180 (2021)
39. Burden, R.L.: Numerical analysis (2011)
40. Atkinson Kendall, E., et al.: An Introduction to Numerical Analysis, pp. 37–45. Wiley Inc., New York (1989)
41. Nwaigwe, C.: An unconditionally stable scheme for two-dimensional convection-diffusion-reaction equations (2022). https://www.researchgate.net/profile/Chinedu-Nwaigwe/publication/357606287_An_Unconditionally_Stable_Scheme_for_Two-Dimensional_Convection-Diffusion-Reaction_Equations/links/61d62472e669ee0f5c8a4a7d/An-Unconditionally-Stable-Scheme-for-Two-Dimensional-Convection-Diffusion-Reaction-Equations.pdf. (Accessed on 19-January-2023)
42. Nwaigwe, C., Makinde, O.D.: Finite difference investigation of a polluted non-isothermal non-newtonian porous media flow. Diffus. Foundat. **26**(4), 145–156 (2019)
43. Nwaigwe, C., Ndu, R.I., Weli, A.: Wall motion effects on channel flow with temperature-dependent transport properties. Appl. Math. **9**(3), 162–168 (2019)
44. Weli, A., Nwaigwe, C.: Numerical analyses of channel flow with velocity-dependent suction and nonlinear heat source. J. Interdiscip. Math. **23**(5), 987–1008 (2020)
45. Nwaigwe, C.: Analysis and application of a convergent difference scheme to nonlinear transport in a brinkman flow. Int. J. Numer. Methods Heat Fluid Flow **30**(10), 4453–4473 (2020)
46. Nwaigwe, C., Weli, A., Makinde, O.D.: Computational analysis of porous channel flow with cross-diffusion. Am. J. Comput. Appl. Math. **9**(5), 119–132 (2019)
47. Nwaigwe, C., Weli, A.: Analysis of two finite difference schemes for a channel flow problem. Asian Res. J. Math. **15**, 1–14 (2019)

System Analysis and Optimization for Random Dynamic Signals

Xiaonan Xiao

Abstract Nowadays big data analysis is one of the most demanding frontier technologies. This article develops the information flow research from the regular information space to the wider information measuring space based on the multi-dimension, multi-angle random dynamic information flow research. In stochastic dynamic delivery system, it estimates A kind of general stable multi-dimensional measurable process. The article investigates a lot of precious information in great generalized process. Thus, it makes the optimal estimation of random signals and the optimal recursive filtering equation under this process. Therefore, it presents an efficient mathematic method and a consistent theoretical foundation for more extensive research has been conducted to improve the efficiency of information transmission systems and optimize their management. Because the system analysis of stochastic dynamic signals has strong global search ability and popularization adaptability, the new intelligent optimization algorithm proposed in this paper is becoming one of the most active research directions in intelligence science, information science and artificial intelligence, and has been rapidly popularized and applied in many engineering fields.

Keywords Generalized process · Random signal · Big date system analysis · Intelligent optimization algorithms · Optimal filtering · Optimization modeling

1 Introduction

Recently the algorithm of partial data and the statistical inference is hot subject. Statisticians are interested in it. They ask for further study. For information transfer optimization, the information system search and analysis as well as simulation and control are a very complex subject. Analysis plays a critical role to improve the efficiency of information decision and enhance the signal transfer [1–3]. This article

X. Xiao (✉)
Xiamen University Tan Kah Kee College, Zhangzhou, Fujian, China
e-mail: xiaoxn@xujc.com

W. Wang et al. (eds.), *Computational Statistics and Data Intelligence*, Springer Proceedings in Mathematics & Statistics 463,
https://doi.org/10.1007/978-981-97-4438-1_10

takes on stochastic recursive filtering method and conducts random optimization analysis. Therefore, it efficiently helps to control the random signal of A kind of general stable measurable multi-dimensional process and helps to solve the optimization modeling. It takes on a thorough study in the optimal filtering [4–7]. In addition, in the generalized process it helps to estimation equation of signals and the optimal filter equation. It greatly efficiently enhances the information transmission system [8–10].

The rise of a new round of global scientific and technological revolution and industrial revolution has led to the accelerated evolution of mathematical technology, led the vigorous development of the digital economy, and had a profound impact on the science and technology, economy and society of various countries, among which intelligent control optimization algorithm and artificial intelligence, as cutting-edge and strategic technologies that lead the future, are having a great impact on the in-depth development of future science and technology. In this regard, more and more scientists firmly believe that intelligent algorithms and artificial intelligence will bring the third technological revolution to human society.

2 Estimation of Random Signals for Generalized Process in Intelligent Decision

Suppose the state of certain "controllable" system is described according to the following equation process $(\xi, \zeta) = [(\xi_1(t), \ldots, \xi_k(t)), (\zeta_1(t), \ldots, \zeta_l(t))]$,

$$t = 0, 1, 2, \ldots, T < \infty$$

$$\xi_{t+1} = a(t)u_t + b(t)\xi_t + c(t)s_1(t+1) \tag{1}$$

$$\zeta_{t+1} = A(t)\xi_t + B(t)s_2(t+1)$$

where the matrixes of stage numbers $a(t), b(t), c(t), B(t)$ and A(t) are $(r \times l), (r \times r), (j \times r), (j \times j)$ correspondingly. The elements they have are stable bound function $t = 0, 1, 2, \ldots, T-1$. Independent random sequence $s_1(t) = (s_{11}(t), \ldots, s_{1k}(t))$, $s_2(t) = (s_{21}(t), \ldots, s_{2l}(t)), t = 1, 2, \ldots, T$ is Gaussian sequence and it holds independent components $H_{s_{ij}^1}(t) = 0, H_{s_{ij}^2}(t) = 1$.

ξ_0- Gaussian random vector $H\xi_0 = n, H[(\xi_0 - n_0)(\xi_0 - n_0)^*] = \rho$ is not decided by sequence $s_i(t), i = 1, 2, t = 1, 2, \ldots, T$. Row vector $v_t = [v_1(t, \zeta), \ldots, v_r(t, \zeta)]$ is also contained in (1), where for each $t = 0, 1, 2, \ldots, T-1$, $v_i(t, \zeta)$ is $\mathrm{F}_t^\zeta = \mu\{\omega : \zeta_0, \zeta_1, \ldots, \zeta_t\}$ observable.

Now let us talk about Control $v = (v_0, v_1, \ldots, v_{r-1})$, suppose

$$\sum_{i=1}^{r} HV_i^2(t,\zeta) < \infty, \quad t = 0,\ 1,\ 2, \ldots,\ T-1. \tag{2}$$

Assuming $v = (v_0, v_1, \ldots, v_{T-1})$

$$U(v) = H\left[\sum_{t=0}^{T-1} (\xi_t^* M(t)\xi_t + v_t^* K(t) v_t) + \xi_T^* M(L)\xi_T\right] \tag{3}$$

$$\overline{U}(v) = H\left(\sum_{t=0}^{T} (n_t^v)^* M(t) \mathrm{n}_t^v + \sum_{t=0}^{T-1} v_t^* K(t) v_t\right) \tag{4}$$

and

$$n_{t+1}^v = a(t)v_t + b(t)n_t^v + C(t)\overline{s^v}(t+1). \tag{5}$$

3 Information Search and System Description

Suppose

$$F(t) = \int_{-\pi}^{\pi} e^{i\lambda t} \frac{R_{n-1}(e^{i\lambda})}{S_n(e^{i\lambda})} \eta(\mathrm{d}\lambda) \tag{6}$$

$$E\eta(\mathrm{d}\lambda) = 0,\ E|\eta(\mathrm{d}\lambda)|^2 = \frac{\mathrm{d}\lambda}{2\pi},$$

$$\mathrm{R}_{n-1}(Z) = \sum_{k=0}^{n-1} \mathrm{a}_k Z^k,\ \mathrm{S}_n(Z) = \sum_{k=0}^{n} b_k z_k,\ b_n = 1,\ b_k, a_k \in R^1$$

All solutions for equation $\mathrm{S}_n(Z) = 0$ are inside the unit.
Suppose

$$f_{\mathrm{F}}(\lambda) = \left|\frac{R_{n-1}(e^{i\lambda})}{S_n(e^{i\lambda})}\right|^2 \tag{7}$$

Then from the following $\eta(d\lambda)$

$$\xi(t) = \int_{-\pi}^{\pi} e^{i\lambda(t-1)} \eta(d\lambda) \tag{8}$$

$$E\xi(t) = 0,\ E|\xi(t)|^2 = \int_{-\pi}^{\pi} \frac{d\lambda}{2\pi} = 1$$

thus

$$E\xi(t)\overline{\xi}(s) = \int_{-\pi}^{\pi} e^{i\lambda(t-s)} \frac{1}{2\pi} d\lambda = \zeta(t, s) \tag{9}$$

in which $\zeta(t, s)$ is a Kronecker symbol.

Suppose

$$F_j(t) = \int_{-\pi}^{\pi} e^{i\lambda t} H_j(e^{i\lambda})\eta(d\lambda),\ j = 1, 2, \ldots, n \tag{10}$$

which $\zeta(t, s)$ is

$$H_j(Z) = Z^{-(n-j)} H_n(Z) + \sum_{k=j}^{n-1} \beta_k Z^{-(k-j+1)},\ j = 1, 2, \ldots, n-1 \tag{11}$$

$$H_n(Z) = -Z^{-1} \sum_{k=0}^{n-1} \alpha_k H_{k+1}(Z) + Z^{-1}\alpha_n \tag{12}$$

and

$$\alpha_1 = a_{n-1}, \quad \alpha_j = a_{n-j} - \sum_{i=1}^{j-1} \alpha_i \beta_{n-j+i}, \quad j = 2, \ldots, n \tag{13}$$

From (11), (12) we can conclude correspondingly.

$$H_j(Z) = Z^{-1}[H_{j+1}(Z) + \alpha_j] \tag{14}$$

$$H_n(Z) = Z^{-1}\left[-\sum_{k=0}^{n-1} \beta_k H_{k+!}(Z) + \alpha_n\right] \tag{15}$$

$$H_n(Z) = Z^{-1}\left\{\sum_{k=0}^{n-1} \beta_k [Z^{-(n-K-1)} H_n(z) + \sum_{j=k+1}^{n-1} \alpha_j Z^{-(j-k)}] + \alpha_n\right\} \tag{16}$$

so

$$H_n(Z) = \frac{R_{n-1}(Z)}{S_n(Z)} \tag{17}$$

From (14) to (17) let us deduce that

$$H_j(Z) = \frac{R_{n-1}(Z)}{S_n(Z)} \tag{18}$$

and according to (8), we have

$$R_{n-1}^{(1)}(z) \equiv R_{n-1}(Z) \tag{19}$$

so $F_1(t) = F(t)$.

4 Optimal Algorithm Analysis

The generalize reliable process $F(t), t = 0, \pm1, \pm2, \ldots$, of the spectrum expression (6) is a constituent of the reliable generalized n-dimensional process $(F_1(t), \ldots, F_n(t))$ which is affiliated with the following recursion equations, $F_1(t) = F(t)$.

$$F_j(t+1) = F_{j+1}(t) + \alpha_j \xi(t+1), j = 1, 2, \ldots, n-1,$$

$$F_n(t+1) = -\sum_{j=0}^{n-1} \beta_j F_{j+1}(t) + \alpha_n \xi(t+1). \tag{20}$$

The process $\xi(t),\ t = 0,\ \pm1,\ \pm2, \ldots$, can be represented by (8)

$$EF_j(s)\xi(t) = 0,\ s < t,\ j = 1, 2, \ldots, n \tag{21}$$

although the coefficient $\alpha_1, \alpha_2, \ldots, \alpha_n$ can be expressed by the Eq. (13).

If $F(t)$, $t = 0$, ±1, $\pm2, \ldots$, is a real process, then every process $\xi(t), F_2(t), \ldots, F_n(t)$ is a real one.

The real reliable k + 1 dimensional process, we can deduce

$$U_t = \int_{-\pi}^{\pi} e^{i\lambda t} H(e^{i\lambda}) \eta(d\lambda) \tag{22}$$

in which $H(Z) = \|H_{r,q}(Z)\|$ is N × m order matrix, N = k + 1. We get the fractional rational element

$$\mathrm{H}_{r,q}(Z) = \frac{R^{(r,q)}_{n_{r,q-1}}(Z)}{S^{(r,q)}_{n_{r,q}}(Z)}. \tag{23}$$

And $\mathrm{H}(\mathrm{d}\lambda) = [\eta_1(\mathrm{d}\lambda), \ldots, \eta_m(\mathrm{d}\lambda)]$ with non-relational components $E\eta_i(\mathrm{d}\lambda) = 0$, $E\left|\eta_j(\mathrm{d}\lambda)\right|^2 = \frac{1}{2\pi}\mathrm{d}\lambda$ supposing the root of the equation $S^{(r,q)}_{n_{r,q}}(Z) = 0$ is within the unit.

For every application process (6),

$$\mathrm{U}_{p,r,q}(t) = \int_{-\pi}^{\pi} e^{i\lambda t} H_{r,q}(e^{i\lambda})\eta_p(\mathrm{d}\lambda) \tag{24}$$

$$\hat{\theta}_{t+1} = b_1\hat{\theta}_t + b_2\zeta_t + a\xi(t+1)$$

$$\zeta_{t+1} = A_1\hat{\theta}_t + A_2\xi_t + B\zeta(t+1) \tag{25}$$

wherein $\xi(t) = (\xi_1(t), \ldots, \xi_m(t))$ is a sequence of non-relational vector for the irrelevant element $E\xi_j(t) = 0,\ E\xi_j^2(t) = 1$

$$\xi_j(t) = \int_{-\pi}^{\pi} e^{i\lambda(t-1)}\eta_j(\mathrm{d}\lambda) \tag{26}$$

and the matrix b_i, B_i, a and A, i = 1, 2, can be immediately deduced.

If v_t, $t = 0, 1, \ldots$, is a Gauss process, $\hat{m}_t = E(\hat{\theta}_t | F_t^{\zeta})$ and $\hat{Y}_t = ([\hat{\theta}_t - \hat{m}_t][\hat{\theta}_t - \hat{m}_t]^*)$ is decided by the following equation:

$$\hat{m}_{t+1} = b_1\hat{m}_t + b_2\xi_t + (aA^* + b_1\hat{r}_t B^*)(AA^* + B_1\hat{r}_1 B^*)^+(\zeta_{t+1} - B_1\hat{m}_t - B_2\zeta_t) \tag{27}$$

$$\hat{r}_{t+1} = b_1\hat{r}_t b_1^* + aa^* - (aA^* + b_1\hat{r}_t B_1^*)(AA^* + B_1\hat{r}_1 B_1^*)(aA^* + b_1\hat{r}_t B_1^*)^*. \tag{28}$$

We can solve this equation by its originated condition: $\hat{m}_0 = E(\hat{\theta}_0|\zeta_0)$, $\hat{r}_0 = E[\hat{\theta}_0 - \hat{m}_0][\hat{\theta}_0 - \zeta_0]^*$.

According to normal correlation theorem

$$\hat{m}_0 = \mathrm{cov}(\hat{\theta}_0, \zeta_0)\mathrm{cov}^+(\zeta_0, \zeta_0)\zeta_0 \tag{29}$$

$$\hat{r}_0 = \mathrm{cov}(\hat{\theta}_0, \hat{\theta}_0) - \mathrm{cov}(\hat{\theta}_0, \zeta_0)\mathrm{cov}^+(\zeta_0, \zeta_0)\mathrm{cov}(\hat{\theta}_0, \zeta_0) \tag{30}$$

Since $\hat{m}_t = E(\hat{\theta}_t|F_t^{\zeta})$ relies on $\zeta_0, \ldots, \zeta_t$, giving $v_t = [\theta_t, \zeta_t]$, optimal linear containing θ_t can be solved by $\zeta_0, \ldots, \zeta_t$. We can get the solution through the Eqs. (27) and (28).

Assuming that (β, α) is a random vector related to $E(\beta^2 + \alpha^2) < \infty$ and $(\tilde{\beta}, \tilde{\alpha})$ is a Gauss vector which has the two identical moments of n-th order as (β, α), $E\tilde{\beta}^i = E\beta^i, E\tilde{\alpha}^i = E\alpha^i, i = 1, 2, E\tilde{\beta}\tilde{\alpha} = E\beta\alpha$. If $l(a)$ is $P_{-\Pi \cdot H}$, we get

$$l(\tilde{\alpha}) = E(\tilde{\beta}|\tilde{\alpha}) \tag{31}$$

the linear function of $a \in R^1$. Then $l(\alpha)$ is optimal linear estimation of the quantity by β and α.

Once the vector is β and α, which is, $\beta = (\beta_1, \ldots, \beta_l), \quad \alpha = (\alpha_1, \ldots, \alpha_k)$, we say the above deduction is still true.

If ζ_t, $t = 0, \pm 1, \pm 2, \ldots$ is a common reliable process with $E\zeta_t \equiv 0$ and assuming spectral density is $\tilde{f}(\lambda) = \left|e^{i\lambda} + 1\right|^2 / \left|e^{2i\lambda} + e^{i\lambda}/2 + 1/2\right|^2$, with the purpose of acquiring ζ by the best linear estimation (which $s \leq t$) of $\zeta_0^s = \{\zeta_0, \ldots, \zeta_s\}$ in the mean square sense, we need to build ζ_t, $t = 0, \pm 1, \pm 2, \ldots$ which has $E\zeta_t \equiv 0$ and a spectral density $f(\lambda) \equiv \tilde{f}(\lambda)$. The process can be computed by

$$\zeta_{t+2} + \frac{1}{2}(\zeta_{t+1} + \zeta_t) = \xi(t+2) + \xi(t+1).$$

Here, $\xi(t), t = 0, \pm 1, \pm 2, \ldots$ random sequence

$$m_{s+1} = -\frac{1}{2}m_s - \frac{1}{2}\zeta_t + \frac{1 - r_s}{2(1 + r_s)}(\zeta_{s+1} - m_s) \tag{32}$$

$$r_{s+1} = \frac{r_s}{1 + r_s} \tag{33}$$

We can easily prove that $m_0 = 0, r_0 = 1$.

Thus we can deduce that the optimal linear estimation reliable sequence ζ_t, $t = 0, \pm 1, \pm 2, \ldots$, is predicated on (32)–(33) as ζ_t of (32) has to be replaced by ζ_t.

5 Application Analysis

Supposing

$$f_\theta(\lambda) = 1/\left|e^{i\lambda} + c_1\right|^2, \; f_\xi(\lambda) = 1/\left|e^{i\lambda} + c_2\right|^2 |c_i| < 1, \quad i = 1, 2$$

wherein, $|C_i| < 1, \quad i = 1, 2$

Supposing the observation process is

$$\zeta_{t+1} = \theta_{t+1} + \zeta_{t+1} = (c_1 - c_2)\theta_t + c_2\xi_t + \varepsilon_1(t+1) + \varepsilon_2(t+1) \tag{34}$$

$$\theta_{t+1} = C_1\theta_t + \xi_1(t+1),\ \xi_{t+1} = C_2\zeta_t + \xi_2(t+1) \tag{35}$$

by taking consideration of (34) and (35) we conclude $\zeta_{t+1} = \theta_{t+1} + \xi_{t+1} = (c_1 - c_2)\theta_t + c_2\xi_t + \xi_1(t+1) + \xi_2(t+1)$.

Therefore, the "unobservable" process θ_t and "observable" process ξ_t satisfy $\theta_{t+1} = C_1\theta_t + \xi_1(t+1)$.

$$\zeta_{t+1} = (c_1 - c_2)\theta_t + c_2\zeta_t + \zeta_1(t+1) + \zeta_2(t+1). \tag{36}$$

According to (22) and (23), the best linear estimate m_t, $t = 0, 1, 2, \ldots$ of θ_t and the filtering MSE $r_t = E(\theta_t - m_t)^2$ satisfy the recursive equation

$$m_{t+1} = c_1 m_t + \frac{2 + (c_1 - c_2)r_t}{5 + (c_1 - c_2)^2 r_t}[\zeta_{t+1} - c_1 - m_t - c_2\zeta_t] \tag{37}$$

$$r_{t+1} = c_1^2 r_t + 1 - \frac{[2 + (c_1 - c_2)r_t]^2}{5 + (c_1 - c_2)^2 r_t} \tag{38}$$

Since the process (θ_t, ζ_t), $t = 0, \pm 1, \pm 2, \ldots$ is a process and we have $E\theta_t = E\zeta_t = 0$ and covariance $d_{11} = E\theta_t^2,\ d_{12} = E\theta_t\zeta_t,\ d_{22} = E\zeta_t^2$, and they all satisfy the equations:

$$\begin{aligned} d_{11} &= c_1^2 d_{11} + 1, \\ d_{12} &= (c_1 - c_2)d_{11} + c_1c_2d_{12} + 3, \\ d_{22} &= (c_1 - c_2)^2 d_{11} + c_2^2 d_{22} + (c_1 - c_2)d_{12} + 4 \end{aligned}$$

we deduce

$$d_{11} = \frac{2}{1 - c_1^2},\ d_{12} = \frac{3}{1 - c_1^2},\ d_{22} = \frac{4 - c_1^2 - c_2^2}{(1 - c_1^2)(1 - c_2^2)}.$$

If we notice (29) and (30), we can conclude the originated conditions

$$\begin{aligned} m_0 &= \frac{d_{12}}{d_{22}}\zeta_0 = \frac{3 - 3c_2^2}{4 - c_1^2 - c_2^2}\zeta_0, \\ r_0 &= d_{11} - \frac{d_{12}^2}{d_{22}} = \frac{2}{3 - c_1^2 - c_2^2} \end{aligned}$$

So "the useful signal" θ_t is deduced through the Eqs. (37) and (38) by the optimal m_t and MSE r_t in the meaning of square of $\zeta_0, \ldots, \zeta_t$. Thus, the problem is solved is based on the originated conditions.

$$m_0 = \frac{3 - 3c_2^2}{4 - c_1^2 - c_2^2}\xi_0,\ r_0 = \frac{2}{3 - c_1^2 - c_2^2}$$

If the target is the parameter, θ_t the result will not change dramatically if we estimate it through the $(\zeta_{-N}, \ldots, \zeta_0, \ldots, \zeta_t)$. Meanwhile, the Eqs. (37) and (38) are still correct, and

$$m_{-N} = \frac{3 - 3c_2^2}{4 - c_1^2 - c_2^2}\zeta_{-N},\ r_{-N} = \frac{2}{3 - c_1^2 - c_2^2}.$$

At present, with the rapid development of the discipline of time series analysis, the importance of stochastic process theory for the establishment, prediction and control analysis of dynamic stochastic data models have been increasingly highlighted.

With the continuous and rapid development of the national economy, the dynamic random data modeling, prediction and control analysis of the production of the vast number of enterprises in the information development decision-making area has increasingly demonstrated its importance.

The sequence of 70 sequential outputs in the production process of a large petrochemical enterprise is shown in Table 1.

The following is a theoretical and technical means of stochastic process and time series analysis to determine the parameters, parameter estimation and model estimation of this sequence model, and test whether its fitting model is consistent with the actual data.

For this purpose, it can be calculated according to the recursive formula, and then the analog image of its autocorrelation function sequence $\{\hat{\rho}_k(\varepsilon),\ k = 1, 2, \ldots, 66\}$ can be obtained, as shown in Fig. 1.

It can be seen that the absolute value of 63 out of 66 numbers is less than $\frac{1}{\sqrt{68}} = 0.12$, and the proportion is $\frac{63}{68} = 0.93$, which is much greater than the probability value of 0.683. Therefore, it can be judged that this residual is classified as an independent sequence, that is, the actual data can accept the above fitting model.

Table 1 Sequential sequence B (ton) of production process output

Group	Z1	Z2	Z3	Z4	Z5	Z6	Z7	Z8	Z9
1–9	47	64	23	71	38	64	55	41	59
10–18	48	71	35	57	40	58	44	80	55
19–27	37	74	51	57	50	60	45	57	50
28–36	45	25	59	50	71	56	74	50	58
37–45	45	54	36	54	48	55	45	57	50
46–54	62	44	64	43	52	38	59	55	41
55–63	53	49	37	35	54	45	68	38	50
64–70	60	39	59	40	57	54	23		

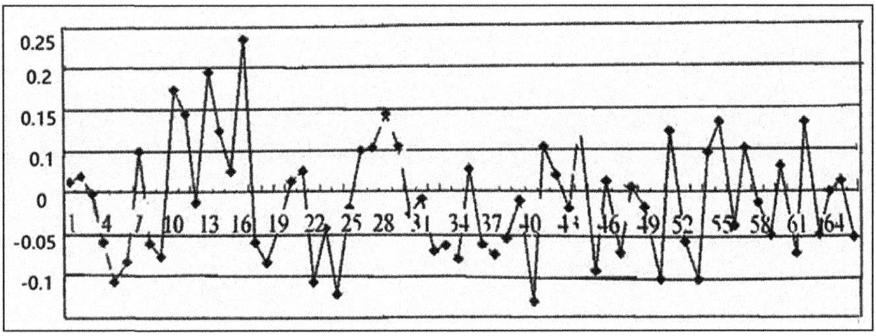

Fig. 1 Sequence of autocorrelation functions

6 Future Work

Big data system analysis and intelligent computing problems are widely present in scientific research and engineering practice [11–13]. The big data system is the way to solve big data service in many fields today [14, 15]. Aiming at the massive characteristics of network traffic in the random dynamic transmission system, in order to make up for the deficiencies of the network massive traffic prediction model to obtain better prediction results of random information flow, this paper uses random optimization analysis, random signal recursive filtering and big data system analysis. The intelligent optimization algorithm of this kind has carried out in-depth research on a kind of generalized random process and partly observable and obtained the optimal filtering equation and estimation equation for the random signal for future research. The precise prediction and control of these processes and the in-depth improvement of the efficiency of such random signal transmission systems provide a very effective mathematical processing method and intelligent optimization algorithm, which are used in engineering optimization, electronic and communication engineering, computer science, and artificial intelligence. It can also be applied in intelligence, intelligent robots, intelligent transportation, intelligent control, space technology, ocean development and many other scientific research fields [16–21].

References

1. Zhou, Y., Zhu, M., Wang, J., Zhang, Z., Xiang, Y., Zhang, J.: Tri-goal evolution framework for constrained many-objective optimization. IEEE Trans. Syst. Man Cybern. 1–14 (2019) https://doi.org/10.1109/tsmc.2018.2858843
2. Yang, Y., Liu, J., Shi, T., Wang, H.: A multi-objective differential evolutionary algorithm for constrained multi-objective optimization problems with low feasible ratio. Appl. Soft Comput. **80**, 42–56 (2019). https://doi.org/10.1016/j.asoc.2019.02.041
3. Long, J.: Growth of solutions of second order complex linear differential equations with entire coefficients. Filomat **32**(1), 275–284 (2018). https://doi.org/10.2298/fil18012751

4. He, F.-L., Chen, W.-N., Hu, X.-M.: Differential evolution with double-level archives for bi-objective cloud task scheduling. In: 2016 IEEE Congress on Evolutionary Computation (CEC), pp. 2942–2949. IEEE (2016). https://doi.org/10.1109/CEC.2016.7744161
5. Peng, H., Ruan, L., Xiang, J.: A note on boundary layer of a nonlinear evolution system with damping and diffusions. J. Math. Anal. Appl. **426**(2), 1099–1129 (2015). https://doi.org/10.1016/j.jmaa.2015.01.053
6. Zou, P., Rajora, M., Liang, S.Y.: A new algorithm based on evolutionary computation for hierarchically coupled constraint optimization: methodology and application to assembly job-shop scheduling. J. Sched. **21**(5), 545–563 (2018). https://doi.org/10.1007/s10951-018-0572-2
7. Liu, B., Fan, R., Liu, H., Zhang, L., Wang, H., Zhang, C.: Learning algorithm of Bayesian network structure based on hybrid Thaliacea differential evolution algorithm. J. Commun. **40**(7), 151–161 (2019). https://doi.org/10.11959/j.issn.1000-436x.2019124
8. Cheng, R., Jin, Y., Olhofer, M., Sendhoff, B.: A reference vector guided evolutionary algorithm for many-objective optimization. IEEE Trans. Evol. Comput. **20**(5), 773–791 (2016). https://doi.org/10.1109/tevc.2016.2519378
9. Dhadwal, M.K., Jung, S.N., Kim, C.J.: Advanced particle swarm assisted genetic algorithm for constrained optimization problems. Comput. Optim. Appl. **58**(3), 781–806 (2014). https://doi.org/10.1007/s10589-014-9637-0
10. Ruan, L., Zhu, C.: Boundary layer for nonlinear evolution equations with damping and diffusion. Discrete Continuous Dyn. Syst. **32**(1), 331–352 (2012). https://doi.org/10.3934/dcds.2012.32.331
11. Zhang, C.-Y., Tu, J.: Growth of solutions to linear differential equations with entire coefficients of slow growth. Electron. J. Diff. Eqs. **2010**, 1–12 (2010). https://ejde.math.txstate.edu/Volumes/2010/43/zhang.pdf
12. Roussas, G.G.: Asymptotic normality of the kernel estimate of a probability density function under association. Stat. Prob. Lett. **50**(1), 1–12 (2000). https://doi.org/10.1016/s0167-7152(00)00072-9
13. Isaacs, I.M.: Finite Group Theory, pp. 1–354. American Mathematical Society, Rhode Island (2011)
14. Kheawhom, S.: Efficient constraint handling scheme for differential evolutionary algorithm in solving chemical engineering optimization problem. J. Indus. Eng. Chem. **16**(4), 620–628 (2010). https://doi.org/10.1016/j.jiec.2010.03.004
15. Lu, L., Yuan, F., Li, R.C.: A new look at the doubling algorithm for a structured palindromic quadratic eigenvalue problem. Numer. Linear Algebra Appl. (2014). https://doi.org/10.1002/nla.1962
16. Xue, L.: Well-posedness and zero micro-rotation viscosity limit of the 2D micropolar fluid equations. Math. Meth. Appl. Sci. **34**, 1760–1777 (2011). https://doi.org/10.1002/mma.1491
17. Yang, Y., Yang, Q.: Singular values of nonnegative rectangular tensors. Front. Math. China **6**(2), 363–378 (2011). https://doi.org/10.1007/s11464-011-0108-y
18. Li, R.-C., Lin, W.-W., Wang, C.-S.: Structured backward error for palindromic polynomial eigenvalue problems. Numerische Mathematik **116**(1), 95122 (2010). https://doi.org/10.1007/s00211-010-0297-4
19. Wei, Z., Zhang, H., Chen, L.: Web anomaly detection model using smotetomek and lightGBM algorithm. J. Chin. Comput. Syst. **3**, 587–592 (2020). CNKI:SUN:XXWX.0.2020-03-025.
20. Liang, J., Chen, J., Zhang, X., Zhou, Y., Lin, J.: One-hot encoding and convolutional neural network based anomaly detection. J. Tsinghua Univ. (Sci. Technol.) **59**(7), 523–529 (2019). https://doi.org/10.16511/j.cnki.qhdxxb.2018.25.061
21. Gu, T., Xu, G., Li, W., Li, J., Wang, Z., Luo, J.: Intelligent house price evaluation model based on ensemble lightGBM and Bayesian optimization strategy. J. Comput. Appl. **361**(9), 290–295 (2020). https://doi.org/10.11772/j.issn.1001-9081.2019122249

Neighbor-Distinguishing Arc Colorings of Several Digraphs

Qiang Ji and Jihui Wang

Abstract If there is a choice function $f{:}A(D) \rightarrow [k]$ such that any arc with a common starting vertex and any arc with a common ending vertex do not have the same color in digraph D, then f is a proper arc coloring. Let $S^-(u)$ and $S^+(u)$ denote by the color sets of the incoming arcs and the outgoing arcs of u separately. If there is a proper arc coloring f such that the color sets of the outgoing arcs of u are not as same as the color sets of the incoming arcs of v for every arc $\overrightarrow{uv} \in A(D)$, then f is neighbour-distinguishing. The neighbor-distinguishing arc coloring f of D requires the minimum arc chromatic number to be denoted as $\chi'_{+,-}(D)$. In this article, we prove the upper bound of a general digraph and the exact results of neighbor-distinguishing arc chromatic number of several special digraph.

Keywords Neighbor-distinguishing arc coloring · General digraph · Special digraph

1 Introduction

We study simple digraphs in this paper. For a digraph $D = (V(D), A(D))$, $d^-(u)$ and $d^+(u)$ is the in-degree and the out-degree of vertex u in digraph D. $\Delta^-(D)$ is the maximum in-degree of D and $\Delta^+(D)$ is the maximum out-degree of D, and let $\Delta^*(D) = \max\{\Delta^-(D), \Delta^+(D)\}$ [1].

If there is a choice function $f : A(D) \rightarrow [k]$ such that any arc with a common starting vertex and any arc with a common ending vertex do not have the same color in digraph D, then f is a proper arc coloring. The proper arc coloring f of digraph D requires the minimum arc chromatic number to be denoted as $\chi'(D)$. Let $S^-(u)$ and $S^+(u)$ denote by the color sets of the incoming arcs and the outgoing arcs of u separately.

Q. Ji · J. Wang (✉)
School of Mathematical Sciences, University of Jinan, Jinan, China
e-mail: ss_wangjh@ujn.edu.cn

W. Wang et al. (eds.), *Computational Statistics and Data Intelligence*, Springer Proceedings in Mathematics & Statistics 463,
https://doi.org/10.1007/978-981-97-4438-1_11

Sopena and Wozniak [2] introduced and investigated the notion of neighbor-distinguishing arc colorings of digraphs in 2019. There is a proper arc coloring f in digraph D, if it satisfies $S^+(u) \neq S^+(v)$ or $S^-(u) \neq S^-(v)$ for every two adjacent vertices u and v, then such arc coloring f is termed as the neighbor-distinguishing.

We generalize the neighbor-distinguishing arc coloring of a digraph and study the neighbour-distinguishing arc colorings of digraphs under $(+, -)$ deformation in this article. There is a proper arc coloring f in digraph D, if it satisfies that the color sets of the outgoing arcs of u are not as same as the color sets of incoming arcs of v for every $\overrightarrow{uv} \in A(D)$, i.e. $S^+(u) \neq S^-(v)$, then f is neighbor-distinguishing. The neighbor-distinguishing arc coloring f of digraph D requires the minimum arc chromatic number to be denoted as $\chi'_{+,-}(D)$.

If there is an arc $\overrightarrow{uv} \in A(D)$ in digraph D, it satisfies $d^+(u) = d^-(v) = 1$, then arc $\overrightarrow{uv}$ is noted as lonely-arc [3]. When a digraph has a lonely-arc, it is not meaningful to consider its neighbor-distinguishing arc coloring. Therefore, all digraphs have at least 3 vertices and do not contain lonely-arc in this paper.

Proposition 1.1 is obtained from the definition of neighbor-distinguishing arc coloring.

Proposition 1.1 *For a digraph D, we have $\chi_{+,-'}(D) \geq \Delta^*(D)$. However, if there is an arc $\overrightarrow{uv} \in A(D)$ such that $d^+(u) = d^-(v) = \Delta^*(D)$, then $\chi_{+,-'}(D) \geq \Delta^*(D) + 1$.*

2 A General Digraph

For every digraph D, we define the corresponding balanced bipartite graph is.

$$B^D = (X,\ Y;\ E),\ X = Y = V(D),\quad E(B^D) = \{uv : u \in X,\ v \in Y,\ \overrightarrow{uv} \in A(D)\},$$

$\Delta^X(B^D)$ and $\Delta^Y(B^D)$ are called the maximum degree of vertices in X and the maximum degree of vertices in Y respectively, then

$$\Delta^-(G) = \Delta^Y(B^G),\ \Delta^+(G) = \Delta^X(B^G),\ \Delta(B^D) = \max\{\Delta^X(B^D),\ \Delta^Y(B^D)\}.$$

And let $\Delta^*(D) = \Delta(B^D) = \max\{\Delta^X(B^D),\ \Delta^Y(B^D)\}.\Delta^*(D) = \Delta(B^D) = \max\{\Delta^X(B^D),\ \Delta^Y(B^D)\}$.

We know that the proper arc coloring f of digraph D is corresponding to the proper edge coloring c of the corresponding balanced bipartite graph B^D. Therefore, Lemma 2.1 is given as follows obviously.

Lemma 2.1 *Let D be a digraph, then $\chi'(D) = \chi'(B^D) = \Delta(B^D) = \Delta^*(D)$.*

The color sets of the edge associated with vertex u is noted as $S(u)$. There is a proper edge coloring c of B, if it satisfies $S(u) \neq S(v)$ for every two adjacent vertices $u \in X$ and $v \in Y$, then such coloring c is called the adjacent vertex distinguishing.

The adjacent vertex distinguishing coloring c of bipartite graph B requires the minimum arc chromatic number to be noted as $\chi_{a'}(B)$ [4].

We clearly observe that the arc coloring f of digraph D is neighbor-distinguishing if and only if the edge coloring c of the corresponding bipartite graph B^D is adjacent vertex distinguishing, so Lemma 2.2 is given.

Lemma 2.2 *Let D be a digraph, then* $\chi_{+,-'}(D) = \chi_{a'}(B^D)$.

Lemma 2.3 ([5]) *If B is a bipartite graph, then* $\chi_{a'}(B) \leq \Delta(B) + 2$.

According to Lemma 2.3, we have $\chi_{+,-'}(D) = \chi_{a'}(B^D) \leq \Delta(B^D)+2 = \Delta^*(D)+2$.

Therefore, we obtain the upper bound of the neighbor-distinguishing chromatic number of a general digraph.

Theorem 2.1 *Let D be a digraph, then* $\chi_{+,-'}(D) \leq \Delta^*(D) + 2$.

3 Special Digraph

Below we study the the exact results of neighbor-distinguishing arc chromatic number of the product digraph between directed path and cycle.

3.1 Cartesian Product Digraph

Let D_1 and D_2 be digraphs without common vertex, $D_1 \circ D_2$ is called the cartesian product digraph, if $V(D_1 \circ D_2) = V_1 \times V_2 = \left\{(u_i, v_j)|u_i \in V_1,\ v_j \in V_2\right\}$ and $A(D_1 \circ D_2) = \left\{\overrightarrow{(u_i, v_j)(u_{i'}, v_{j'})}|u_i = u_{i'},\ \overrightarrow{v_j v_{j'}} \in A_2 \, or \, v_j = v_{j'},\ \overrightarrow{u_i u_{i'}} \in A_1\right\}$ [6].

The neighbor-distinguishing arc chromatic number of cartesian product digraphs have the following results.

Theorem 3.1 *Let* $m \geq 3$, $n \geq 3$, *and* $m \in N^*$, $n \in N^*$, *then*

$$\chi_{+,-'}(P_m \circ P_n) = \Delta^*(P_m \circ P_n) + 1.$$

Proof Let $P_m = u_1 u_2 \cdots u_m$, $P_n = v_1 v_2 \cdots v_n$. Based on the structure of Cartesian product digraph, we have $\Delta^*(P_m \circ P_n) = 2$. By Proposition 1.1, we have $\chi_{+,-'}(P_m \circ P_n) \geq 3$. In order to prove $\chi_{+,-'}(P_m \circ P_n) \leq 3$, we define an arc coloring f of $P_m \circ P_n$ using three colors $\{0, 1, 2\}$. Let

$$f\left(\overrightarrow{(u_i, v_j)(u_i, v_{j+1})}\right) = i(\text{mod}3), 1 \leq j \leq n - 1,$$

$$f\left(\overrightarrow{(u_i, v_j)(u_{i+1}, v_j)}\right) = i + 2(\text{mod}3),\ 1 \le i \le m - 1.$$

For every $(u_i, v_j) \in V(P_m \circ P_n)$, the color of the incoming arcs of (u_i, v_j) is different and the color of outgoing arcs of (u_i, v_j) is different, so f is a proper arc coloring of $P_m \circ P_n$.

Below we prove that f is neighbor-distinguishing arc coloring of $P_m \circ P_n$.

(1) For every arc on the boundary, since the out-degree of the start vertex is different from the in-degree of the terminal vertex on the boundary, the adjacent vertex can be distinguished.

(2) For every $\overrightarrow{(u_i, v_j)(u_{i\prime}, v_{j\prime})} \in A(P_m \circ P_n)$ in the interior of $P_m \circ P_n$, satisfying $i\prime = i$ or $i\prime = i + 1$. $\overrightarrow{(u_i, v_j)(u_i, v_{j+1})}$ and $\overrightarrow{(u_i, v_j)(u_{i+1}, v_j)}$ are the outgoing arcs of (u_i, v_j), $\overrightarrow{(u_{i\prime-1}, v_{j\prime})(u_{i\prime}, v_{j\prime})}$ and $\overrightarrow{(u_{i\prime}, v_{j\prime-1})(u_{i\prime}, v_{j\prime})}$ are the incoming arcs of $(u_{i\prime}, v_{j\prime})$, under the coloring f,

$$f\left(\overrightarrow{(u_i, v_j)(u_i, v_{j+1})}\right) = i(\text{mod}3),\quad f\left(\overrightarrow{(u_i, v_j)(u_{i+1}, v_j)}\right) = i + 2(\text{mod}3),$$
$$f\left(\overrightarrow{(u_{i\prime-1}, v_{j\prime})(u_{i\prime}, v_{j\prime})}\right) = i + 1(\text{mod}3),\quad f\left(\overrightarrow{(u_{i\prime}, v_{j\prime-1})(u_{i\prime}, v_{j\prime})}\right) = i(\text{mod}3),$$

so $S^+((u_i, v_j)) \neq S^-((u_{i\prime}, v_{j\prime}))$.

Therefore, $\chi'_{+,-}(P_m \circ P_n) = \Delta^*(P_m \circ P_n) + 1$.

Theorem 3.2 *Let $m \ge 3$, $n \ge 3$, and $m \in N^*$, $n \in N^*$, then*
$\chi_{+,-\prime}(P_m \circ C_n) = \Delta^*(P_m \circ C_n) + 1$.

Proof Based on the structure characteristics of digraph $P_m \circ C_n$, we obtain cartesian product digraph $P_m \circ C_n$ is adding $\overrightarrow{(u_i, v_n)(u_i, v_1)}$ to $P_m \circ P_n$. Next, we extend the f of $P_m \circ P_n$ to neighbor-distinguishing arc coloring of $P_m \circ C_n$.

Let $f\left(\overrightarrow{(u_i, v_n)(u_i, v_1)}\right) = i(\text{mod}3)$. Except $\overrightarrow{(u_i, v_n)(u_i, v_1)}$, other verices are still distinguishable from adjacent vertex. We only prove the color sets of outgoing arcs of (u_i, v_n) are different from the incoming arcs of (u_i, v_1). $\overrightarrow{(u_i, v_n)(u_i, v_1)}$ and $\overrightarrow{(u_i, v_n)(u_{i+1}, v_n)}$ are the outgoing arcs of (u_i, v_n), $\overrightarrow{(u_i, v_n)(u_i, v_1)}$ and $\overrightarrow{(u_{i-1}, v_1)(u_i, v_1)}$ are the incoming arcs of (u_i, v_1), under the coloring f,

$$f\left(\overrightarrow{(u_i, v_n)(u_i, v_1)}\right) = i(\text{mod}3),\quad f\left(\overrightarrow{(u_i, v_n)(u_{i+1}, v_n)}\right) = i + 2(\text{mod}3)$$
$$f\left(\overrightarrow{(u_{i-1}, v_1)(u_i, v_1)}\right) = i + 1(\text{mod}3),$$

thus $S^+((u_i, v_n)) \neq S^-((u_i, v_1))$. Therefore, f is neighbor-distinguishing arc coloring of $P_m \circ C_n$, so $\chi_{+,-\prime}(P_m \circ C_n) = \Delta^*(P_m \circ C_n) + 1$.

3.2 Strong Product Digraph

Let D_1 and D_2 be digraphs without common vertex, $D_1 \otimes D_2$ is called the strong product digraph of D_1 and D_2, if $V(D_1 \otimes D_2) = V_1 \times V_2 = \left\{(u_i, v_j)|u_i \in V_1,\ v_j \in V_2\right\}$ and $A(D_1 \otimes D_2) = A(D_1 \circ D_2) \cup \left\{\overrightarrow{(u_i, v_j)(u_{i'}, v_{j'})}|\overrightarrow{u_iu_{i'}} \in A_1 and \overrightarrow{v_jv_{j'}} \in A_2\right\}$.

The neighbor-distinguishing arc chromatic number of strong product digraphs have the following results.

Theorem 3.3 *Let $m \geq 3$, $n \geq 3$, and $m \in N^*$, $n \in N^*$, then*

$$\chi_{+,-'}(P_m \otimes P_n) = \Delta^*(P_m \otimes P_n) + 1.$$

Proof Let $P_m = u_1u_2 \cdots u_m$, $P_n = v_1v_2 \cdots v_n$. We have $\Delta^*(P_m \otimes P_n) = 3$. By Proposition 1.1, we have $\chi_{+,-'}(P_m \otimes P_n) \geq 4$. In order to prove $\chi_{+,-'}(P_m \otimes P_n) \leq 4$, we define an arc coloring f of $P_m \otimes P_n$ using four colors $\{0, 1, 2, 3\}$. Let

$$f\left(\overrightarrow{(u_i, v_j)(u_i, v_{j+1})}\right) = i(\text{mod}4),\ 1 \leq j \leq n-1,$$
$$f\left(\overrightarrow{(u_i, v_j)(u_{i+1}, v_j)}\right) = i+2(\text{mod}4),\ 1 \leq i \leq m-1,$$
$$f\left(\overrightarrow{(u_i, v_j)(u_{i+1}, v_{j+1})}\right) = i+3(\text{mod}4),\ 1 \leq i \leq m-1,\ 1 \leq j \leq n-1.$$

According to the coloring f, can be obtained f is a proper arc coloring of $P_m \otimes P_n.P_m \otimes P_n$.

Below we prove f is neighbor-distinguishing.

(1) For every arc on the boundary of $P_m \otimes P_n$, since the out-degree of the start vertex is different from the in-degree of the terminal vertex on the boundary, the adjacent vertex can be distinguished.

(2) We prove that every arc in the interior of $P_m \circ P_n$ are neighbor-distinguished.

For $\overrightarrow{(u_i, v_j)(u_{i+1}, v_j)}$, $2 \leq j \leq n-1$, under the coloring f, we have

$$S^+((u_i, v_j)) = \{i(\text{mod}4), i+3(\text{mod}4), i+2(\text{mod}4)\},$$
$$S^-((u_{i+1}, v_j)) = \{i+1(\text{mod}4), i+3(\text{mod}4), i+2(\text{mod}4)\},$$

so the color sets of outgoing arcs of (u_i, v_j) and the incoming arcs of (u_{i+1}, v_j) are different.

It is similarly proved that for $\overrightarrow{(u_i, v_j)(u_i, v_{j+1})}$ and $\overrightarrow{(u_i, v_j)(u_{i+1}, v_{j+1})}$, the color sets of outgoing arcs of the start vertex and the incoming arcs of the terminal vertex are different.

For every $\overrightarrow{(u_i, v_j)(u_{i'}, v_{j'})} \in A(P_m \otimes P_n)$, we have $S^+((u_i, v_j)) \neq S^-((u_{i'}, v_{j'}))$. Therefore, $\chi_{+,-'}(P_m \otimes P_n) = \Delta^*(P_m \otimes P_n) + 1$.

Theorem 3.4 *Let $m \geq 3$, $n \geq 3$, and $m \in N^*$, $n \in N^*$, then*

$$\chi_{+,-'}(P_m \otimes C_n) = \Delta^*(P_m \otimes C_n) + 1.$$

Proof Based on the structure characteristics of digraph $P_m \otimes C_n$, we obtain strong product digraph $P_m \otimes C_n$ is adding $\overrightarrow{(u_i, v_n)(u_i, v_1)}$ and $\overrightarrow{(u_i, v_n)(u_{i+1}, v_1)}$ to $P_m \otimes P_n$.

We define an arc coloring f of $P_m \otimes P_n$ using four colors {0, 1, 2, 3}. Let

$$f\left(\overrightarrow{(u_i, v_j)(u_i, v_{j+1})}\right) = i(\text{mod}4),\ 1 \le j \le n-1,$$
$$f\left(\overrightarrow{(u_i, v_j)(u_{i+1}, v_j)}\right) = i + 2(\text{mod}4),\ 1 \le i \le m-1,$$
$$f\left(\overrightarrow{(u_i, v_j)(u_{i+1}, v_{j+1})}\right) = i + 3(\text{mod}4),\ 1 \le i \le m-1, 1 \le j \le n-1,$$
$$f\left(\overrightarrow{(u_i, v_n)(u_i, v_1)}\right) = i(\text{mod}4),\ f\left(\overrightarrow{(u_i, v_n)(u_{i+1}, v_1)}\right) = i + 3(\text{mod}4).$$

Except (u_i, v_n) and (u_i, v_1), other vertices associated with arcs are still proper arc coloring. We only prove the color of the outgoing arcs of (u_i, v_n) is different and the incoming arcs of (u_i, v_1) is different. Since the outgoing arc colors of (u_i, v_n) are $i(\text{mod}4)$, $i + 2(\text{mod}4)$ and $i + 3(\text{mod}4)$, the incoming arc color of (u_i, v_1) are $i(\text{mod}4)$, $i + 1(\text{mod}4)$ and $i + 2(\text{mod}4)$, so f is a proper arc coloring of $P_m \otimes C_n$.

Below we prove that f is neighbor-distinguishing arc coloring of $P_m \otimes C_n$, we only prove the start vertex of (u_i, v_n) associated arcs are distinguishable from the terminal vertex, and the start vertex of (u_i, v_1) associated arcs are distinguishable from the terminal vertex.

We verify the case where (u_i, v_n) is the start vertex, and other cases are similar. The three types outgoing arcs of (u_i, v_n) are $\overrightarrow{(u_i, v_n)(u_{i+1}, v_n)}$, $\overrightarrow{(u_i, v_n)(u_i, v_1)}$ and $\overrightarrow{(u_i, v_n)(u_{i+1}, v_1)}$, so $S^+((u_i, v_n)) = \{i + 2(\text{mod}4), i(\text{mod}4), i + 3(\text{mod}4)\}$.

(1) For $\overrightarrow{(u_i, v_n)(u_{i+1}, v_n)}$, the color sets of the incoming arcs of (u_{i+1}, v_n) are $S^-((u_{i+1}, v_n)) = \{i + 2(\text{mod}4), i + 3(\text{mod}4), i + 1(\text{mod}4)\}$.

(2) For $\overrightarrow{(u_i, v_n)(u_i, v_1)}$, the color sets of the incoming arcs of (u_i, v_1) are $S^-((u_i, v_n)) = \{i(\text{mod}4), i + 1(\text{mod}4), i + 2(\text{mod}4)\}$.

(3) For $\overrightarrow{(u_i, v_n)(u_{i+1}, v_1)}$, the color sets of the incoming arcs of (u_{i+1}, v_1) are $S^-((u_{i+1}, v_1)) = \{i + 3(\text{mod}4), i + 2(\text{mod}4), i + 1(\text{mod}4)\}$.

Therefore, for every $\overrightarrow{(u_i, v_j)(u_{i'}, v_{j'})} \in A(P_m \otimes C_n)$, the color sets of the outgoing arcs of (u_i, v_j) and the incoming arcs of $(u_{i'}, v_{j'})$ are different, and we have $S^+((u_i, v_j)) \ne S^-((u_{i'}, v_{j'}))$, so $\chi_{+,-'}(P_m \otimes C_n) = \Delta^*(P_m \otimes C_n) + 1$.

4 Conclusion

In this paper, we generalize the neighbor-distinguishing arc colorings of digraphs, and study the neighbour-distinguishing arc colorings of digraphs under $(+, -)$ deformation. We study the neighbor-distinguishing arc chromatic number of several digraphs, and prove the upper bound of a general digraph and the exact results of neighbor-distinguishing arc chromatic number of several special digraph.

Acknowledgements This work is helped by the National Natural Science Foundation of China (12071351) and the Natural Science Foundation of Shandong Provence (ZR2020MA043).

References

1. Bang-Jensen, J., Gregory, Z.G.: Digraphs: Theory, Algorithms and Applications. Springer, Science and Business Media (2008)
2. Sopena, E., Woźniak, M.: A note on the neighbour-distinguishing index of digraphs. Ars Mathematica Contemporanea **23**(1), P1. 07–7 (2013)
3. Bensmail, J., Lyngsie, K.: 1-2-3 Conjecture in digraphs: more results and directions. Discret. Appl. Math. **284**, 124–137 (2020)
4. Zhongfu, Z., Linzhong, L., Jianfang, W.: Adjacent strong edge coloring of graphs. Appl. Math. Lett. **15**(5), 623–626 (2002)
5. Balister, P.N., et al.: Adjacent vertex distinguishing edge-colorings. SIAM J. Discret. Math. **21**(1), 237–250 (2007)
6. Bang-Jensen, J., Gregory, Z.G.: Classes of directed graphs. Springer, Cham (2018)
7. Li, H., et al.: Vertex-distinguishing proper arc colorings of digraphs. Discret. Appl. Math. **209**, 276–286 (2016)
8. Aigner, M., Triesch, E., Tuza, Z.: Irregular assignments and vertex-distinguishing edge-colorings of graphs. Ann. Discret. Math. **52**, 1–9 (1992)
9. Hocquard, H., Montassier, M.: Adjacent vertex-distinguishing edge coloring of graphs with maximum degree Δ. J. Comb. Optim. **26**(1), 152–160 (2013)
10. Edwards, K., Horňák, M., Woźniak, M.: On the neighbour-distinguishing index of a graph. Graphs Combinat. **22**(3), 341–350 (2006)

Modelling Odds and Probability of Stillbirth Due to Maternal Demography (Race, Marital Status, Maternal Age and Antenatal Care) Among Women in Malaysia: Case Study in a Public Hospital at Selangor

Nur Syamimi Binti Haji Abu Bakar, Adriana Binti Ismail, Anita Rina Binti Ali, Nur Aqilah Alias, Norshahida Binti Shaadan, and Halilah Binti Haron

Abstract Stillbirth is a serious problem globally with approximately 2.65 million third trimester stillborn baby cases recorded around the world each year. In Malaysia, there are very limited studies on the risks of stillbirth. Thus, the objective of this study was to investigate the odds and probability of stillbirth due to maternal demography (race, marital status, maternal age and antenatal care) among women in Malaysia. A case control retrospective study was undertaken at one public hospital in Selangor. A descriptive analysis was conducted to know the percentage of mothers with maternal demography with birth of their babies. Chi-square test of independence was conducted to identify the association between maternal demography and stillbirth. After that, the logistic regression is conducted to investigate the odds and probability of stillbirth due to maternal demography (race, marital status, maternal age and antenatal care) among women in Malaysia. A descriptive result of the variable for stillborn babies indicated that mothers who are 'Bumiputera' (66.5%), married (94.2%), age less than or equal to 35 years old (77.4%) and having antenatal care (88.4%). The chi square tests of independence results concluded that only age and antenatal care are associated with birth of baby. Based on the odd ratio it can be conclude that a pregnant woman whose age is greater than 35 years old and with no antenatal care during pregnancy have a higher risk of getting stillbirth baby.

Keywords Chi-square test of independence · Logistic regression

N. S. B. H. A. Bakar (✉) · A. B. Ismail · N. B. Shaadan · H. B. Haron
College of Computing, Informatics and Mathematics, Universiti Teknologi MARA Shah Alam, 40450 Shah Alam, Selangor, Malaysia
e-mail: syamimi@uitm.edu.my

A. R. B. Ali · N. A. Alias
Obstetrics and Gynaecology Department, Hospital Sultan Idris Shah Serdang, Selangor Darul Ehsan, Jalan Puchong, 43000 Kajang, Malaysia

W. Wang et al. (eds.), *Computational Statistics and Data Intelligence*, Springer Proceedings in Mathematics & Statistics 463,
https://doi.org/10.1007/978-981-97-4438-1_12

1 Introduction

Stillbirth can be divided into two types which are macerated stillbirth and fresh stillbirth. Other names used for macerated stillbirth and fresh stillbirth are intrauterine death and intrapartum death. Macerated stillbirth is a death of a foetus with birthweight ($\geq$500 g), gestational age ($\geq$22 weeks) and the presence of peeling of skin. Meanwhile, for fresh stillbirth, there is the absence of peeling of skin [1]. Usually, for macerated stillbirth, the baby may have died during late pregnancy which means the baby might have died in the womb or before giving birth while fresh stillbirth is a baby who may have died during labor or birth [2]. Therefore, the physical of the baby will help in estimating the time of death in stillborn fetus either in the womb or during labor.

A pregnant woman should always notice if there are any changes in the movement of the baby in the womb. If the movement and reaction of the baby in the womb keep reducing, it shows there are some problems with the baby. The main signs of stillbirth are when there is no heartbeat and movement of the baby. Other signs of stillbirth are when the pregnant mother feel cramps, pain, and fluid loss or have bleeding from the vaginal. So, the best way is to meet the doctor, to know the progress or problem of the baby. Frequently, the doctor will check the pregnant mother's health and listen to the baby's heartbeat using heart rate monitor but if there is no sign of a heartbeat, the doctor will do an ultrasound to confirm the death of a baby [3].

After the doctor has already confirmed the death of the baby, the doctor will give some suggestions and options for giving birth. The suitable time and best method for delivery of a stillborn baby depend on the gestational age and medical history. From the discussion with the doctor at the hospital, the stillborn baby can be delivered normally after two weeks of death in the womb. Unfortunately, the doctor will suggest medication to induce labour to avoid a baby to degenerate in the womb and doctor can also easily identify cause of death for stillborn baby. Furthermore, most of stillborn babies are delivered normally compared to caesarean delivery and only in some cases which requires a baby to be delivered by caesarean such as multiple pregnancies because the doctor needs to save the other baby [4].

Stillbirth is a serious issue that needs to be known especially among pregnant women. Pregnant women should always know whether they are at a high or low risk in getting a stillborn baby. During pregnancy, many problems can be encountered with the early preparation and the awareness of the mother. Therefore, it is important to meet with doctors regularly to ensure that the mother and the baby are always in a good condition. When the appointment with the doctor is done regularly, the progress of the baby's growth can be monitored and any problems during pregnancy can be avoided and controlled. So, an alive and healthy baby will be born and the proportion of stillborn baby can be reduced.

Stillbirth is a serious problem globally, with approximately 2.65 million third trimester stillborn baby recorded around the world each year with about 98 percent, occurring in low and medium income countries [5]. The country with the lowest rate of stillbirth babies, which is about 1.3 stillbirths per 1000 total births, is Iceland while Denmark is next at 1.7 per 1000 births. The country with the worst stillbirth rate is Pakistan with 43.1 stillbirths per 1000 total births. Malaysia which stands at rank 54, with stillbirth rate of around 4 per 1000 followed very closely to Indigenous Australians and Colombia [6]. Identifying the risks of stillbirth can help reduce the number of stillbirths. There are very limited studies on the risks of stillbirth in Malaysia.

The objective of the study was to investigate the odds and probability of stillbirth due to maternal demography such as race, marital status, maternal age and antenatal care among women in Malaysia and to examine the association between maternal demography such as race, marital status, maternal age and antenatal care with the birth of baby (alive, stillbirth).

2 Methodology

The study was focused on only one public hospital in Selangor. The data was collected from years 2013 to 2015 based on the list of all patients who delivered their baby at the Obstetrics and Gynaecology Department. The data was focused on the alive and stillbirth baby including macerated and fresh stillbirth.

This was a case control retrospective study. This type of design was chosen because comparison can be made between patients who delivers stillbirth babies (cases) with patients who deliver alive babies (controls) in order to make a retrospective review to compare risk factors in each group.

The design was used because it requires less time to complete because of the presence of the occurrence, in which at the same time, researcher can look at multiple risk factors with lower in cost as compared to a prospective study. Furthermore, a retrospective cohort study also allows the investigator to describe a population over time or obtain preliminary measures of association to develop future studies and interventions.

The samples for stillbirth in the study were 155 cases. On the other hand, the samples collected randomly for alive were 212. So, the total sample size was 367. Secondary data was used in this study and the collected data was basically about maternal demography including race, marital status, maternal age, antenatal care and also on the birth of baby, alive or stillbirth. The maternal demography are the independent variables and birth of baby is a dependent variable. The conceptual framework shows the relationship between independent and dependent variables (Fig. 1).

Descriptive analysis was based on contingency table. In this study, variables for birth of baby and maternal demography (race, marital status, maternal age and antenatal care) are categorical data thus contingency table was used to show the percentage and frequency [7].

Chi-Square analysis was conducted to examine the association between maternal demography such as race, marital status, maternal age and antenatal care towards

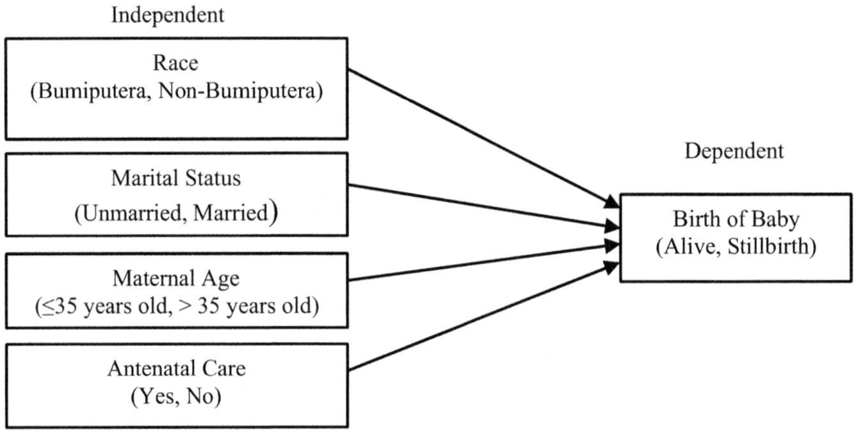

Fig. 1 Conceptual framework

birth of baby (alive and stillbirth). There is an association if the p-value is less than or equal to $\alpha = 0.05$ [9].

After that, logistic regression is conducted to calculate the odds ratio to measure the effect size, describing the strength of association or non-independent between two data values. Basically, the odds ratio was used to measure if there was an association between two variables. Furthermore, the odds ratio for a predictor is defined as the relative amount by which the odd of the outcome increase (OR greater than 1.0) or decrease (OR less than 1.0) when the value of the predictor variable is increased by 1.0 units [8]. Here, odds ratio was used to investigate the association between maternal demography such as race, marital status, maternal age and antenatal care.

3 Results

3.1 Descriptive Analysis

Table 1 shows the descriptive analysis on the maternal demography that relate with a mother that deliver alive and stillbirth baby. The descriptive result of the variable for stillborn babies indicated that majority of the mothers are Bumiputera (66.5%), married had (94.2%), age less than or equal to 35 years old (77.4%) and had antenatal care (88.4%),

While, the result of the alive babies has higher percentage of mothers who are Bumiputera (69.8%), married (95.3%), age less than or equal to 35 years old (88.2%) and had antenatal care (96.2%).

Table 1 Maternal demography

Variable	Alive n (%)	Stillbirth n (%)
Race		
Bumiputera	148 (69.8)	103 (66.5)
Non-Bumiputera	64 (30.2)	52 (33.5)
Marital status		
Unmarried	10(4.7)	9 (5.8)
Married	202 (95.3)	146 (94.2)
Maternal age		
≤35 years	187 (88.2)	120 (77.4)
>35 years	25 (11.8)	35 (22.6)
Antenatal care		
No	8 (3.8)	18 (11.6)
Yes	204 (96.2)	137 (88.4)

Table 2 Chi-square test for associations

Determinants	Chi square	P-value
Race	0.468	0.494
Marital status	0.216	0.642
Age	**7.620**	**0.006**
Antenatal care	**8.359**	**0.004**

3.2 *Test of Independence (Chi Square Test)*

Chi-square analysis for association was conducted to test the relationship or association between the dependent and independent variables. Table 2 shows the Chi-Square results for the four factors of maternal demography. The result shows that only age and antenatal care is associated with stillbirth based (p value less than 0.05). The race and marital status are not associated with stillbirth because the p-value is greater than $\alpha = 0.05$.

3.3 *Logistic Regression (Odd Ratio)*

Based on the chi square test, the logistic regression is applied for the two significance variables age and antenatal care to calculate the odd ratio. Table 3 below shows the results on the odds ratio for maternal demography. Based on the odds ratio, a mother whose age is greater than 35 years old has 2.182 times higher odds (95 CI: 1.244 to 3.827) of getting a stillbirth baby compared to a mother whose age is less than or equal to 35 years old. A mother who has no antenatal care has 3.356 times higher odds (95 CI: 0.126 to 0.706) of getting a stillbirth baby compared to a mother that

Table 3 Odds ratio for antenatal complications

Determinants	Odd ratio	95% confidence interval
Age	2.182	1.244–3.827
Antenatal Care	0.298	0.126–0.706

have antenatal care. Thus, a pregnant woman whose age is greater than 35 years old and with no antenatal care during pregnancy have a higher risk of getting stillbirth baby.

4 Discussion

The result for the races showed higher percentage among Bumiputera with 66.5% compared to others. The result using Chi-Square test showed there is no association between races and birth of a baby. So, race is not a risk of stillbirth.

Most of the stillbirth mothers are married (94.2%), however, marital status is not associated with the birth of baby. Hence, marital status is not a risk of stillbirth. This is supported by another study that showed there is no association between marital status and stillbirth [10].

Maternal age showed higher percentage of stillbirth babies by mothers whose age are less than or equal to 35 years old. The results used Chi-Square test showed the association between maternal age and the birth of baby. The odds ratio also found that a mother with maternal age above 35 years old are more likely to have stillbirth baby compared to a mother with maternal ages less than or equal to 35 years old. Hence, the advanced maternal age has the higher risk of stillbirth despite after controlling for parity, ethnicity, BMI and medical complications [11]. In addition, maternal age older than 35 years was related with an increase of risks of stillbirth and if the age increasing, the higher risk to get stillbirth baby [5]. So, if a woman marries early and gets pregnant at a young age, the risks of stillbirth can be prevented.

The percentage of pregnancy women who had antenatal care was 88.4%. There is a significant association between antenatal cares with the birth of the baby. The odds ratio also shows that antenatal care is one of the risk factors of stillbirth. Most of the past studies showed that a mother who received antenatal care during pregnancy was less likely to have stillbirth baby compared to the mother without antenatal care during pregnancy. Women who lack antenatal care are at risk for stillbirths and inadequate antenatal care was a factor most commonly associated with stillbirth in developing countries [12, 13]. So, antenatal care is associated and is one of the risk factors of stillbirth.

5 Conclusion

The objective of the study was to investigate the odds and probability of stillbirth due to maternal demography (race, marital status, maternal age and antenatal care) among women in Malaysia and to examine the association between maternal demography (race, marital status, maternal age and antenatal care) among women in Malaysia with the birth of baby (alive, stillbirth). The Chi-square test of independent is a statistical method used for testing the relationship between dependent and independent variable. The results showed that the significant variables are maternal age and antenatal care. The odds ratio showed that pregnant woman with age more than 35 years old and with no antenatal care have a higher risk of getting stillbirth baby.

A doctor should always monitor a patient that has a high risk in having a pregnancy problem and pregnant women especially with age more than 35 years old should regularly make check-ups with their doctors. This is to ensure the progress of their baby as well as maternal health will be checked by the doctor to see whether the women are in a good condition or not. The check-up will help to detect any problems during pregnancy. The progress of the baby also can be monitored based on the ultrasound as well as the baby's heartbeat. It is important to make sure that the baby is developing properly and that the placenta is healthy and of normal size. The doctor will also teach ways to monitor baby's movement by calculating the kicking of the baby. This method is useful in order to know the progress of the baby and it is also the best way to avoid a stillbirth baby.

Acknowledgements We would like to express our gratitude and appreciation to Medical Research Ethics Committee (MREC) for the permission to collect data at this public hospital. Thank you to Dr. Anita Rina Ali and Dr. Nur Aqilah Alias for their collaboration in helping with the data collection from the public hospital. We owe our utmost gratitude and appreciation to Universiti Teknologi MARA for this Lestari Grant.

References

1. Reporting Format on Stillbirths & Neonatal Deaths, Accessed from http://www.jknselangor.moh.gov.my/documents/pdf/2013/borang/PerinatalPanduan.pdf (2000)
2. Understanding stillbirth – BabyCenter, Accessed from http://www.babycenter.com.my/a1014800/understanding-stillbirth (2015)
3. Stillbirth March of Dimes, Accessed from http://www.marchofdimes.org/complications/stillbirth.aspx (2015)
4. Sulaiman, S., Othman, S., Razali, N., Hassan, J.: Obstetric and perinatal outcome in teenage pregnancies **19**, 74–77 (2013)
5. Froen, J.F., Lawn, J.E., Heazell, A.E.P., Flenady, V., Bernis, L.D., Kinney, M.V., Leishe, S.H.: 2016 Ending Preventable Stillbirths (2016)
6. Vicki, F.: Stillbirth: Risk Factors and Opportunities for Prevention. University of Queensland, Mater Research Institute (2015)
7. Agresti, A.: An Introduction to Categorical Data Analysis 2nd edn. Wiley (2007)
8. Montgomery, D.C., Peck, E.A., Vining, G.G.: Introduction to linear Regression Analysis, 5th edn. Wiley, Inc. (2012)

9. Triola, M.F.: Elementary Statistics Using EXCEL, 2nd edn (2004)
10. Sutan, R., Campbell, D., Prescott, G., Smith, W.: The risk factors for unexplained antepartum stillbirths in Scotland, 1994 to 2003. J. Perinatol. **30**, 311–318 (2010)
11. Beauclair, R., Petro, G., Myer, L.: The Association Between Timing of Initiation of Antenatal Care and Stillbirths: A Retrospective Cohort Study of Pregnant Women in Cape Town, South Africa, pp. 1–10 (2014)
12. Hassan, H., Sutan, R., Azmi, N.A.M., Ahmad, S., Jaafar, R.: Trend of Stillbirths and Neonatal Deaths in University **3**, 241–248 (2014)
13. Aminu, M., Unkels, R., Mdegela, M., Utz, B., Adaji, S., Broek, V.D.: Causes of and factors associated with stillbirth in low- and middle-income countries: a systematic literature review. J. Obstet. Gynaecol. **121**, 141–153 (2014)

Effectiveness of Roelands Formula with Constant VPC in EHL of Higher Pair Contact

Jiajia Zhang, Chengzhao Liu, and Wei Ding

Abstract An essential issue concerning the elastohydrodynamic lubrication (EHL) problem in the higher pair contacts of gear and bearing is the viscosity calculation of lubricating oil, which has a non-negligible effect on the EHL characteristic parameters such as film thickness, friction coefficient, temperature rise of oil film and substrate and so on. Roelands empirical formula is commonly used in the calculation of lubricant viscosity, and considered to have higher accuracy than exponential function. However, the operation temperature and pressure of the lubricating oil have exceeded its application range in the contact of the high pair. In this paper, the effectiveness of Roelands formula with constant VPC (viscosity-pressure coefficient) in EHL analysis of higher pairs is investigated. The film thickness and temperature rise of EHL point contact are computed according to the fitted VPC expression varying with the temperature, and the results are compared with those with the constant VPC. The results show that the lubricant viscosity, film thickness and film temperature rise of EHL are highly sensitive to the VPC at low temperature and high pressure. Even a modest change in the VPC produces an important rise/reduction in the above characteristic parameters. The Roelands formula with constant VPC causes a certain deviation to the central film thickness (CFT) and maximum film temperature (MFT) in the EHL point contact when the reference temperature is small, and the deviation decreases gradually with the increasing of the reference temperature.

Keywords EHL · Roelands formula · Higher pair contact · Viscosity-pressure coefficient

J. Zhang (✉)
Research Institute CSSC Haizhuang Windpower Co., Ltd., Chongqing, China
e-mail: jjiazhang@cqu.edu.cn

C. Liu · W. Ding
SKLMT, Chongqing University, Chongqing, China

W. Wang et al. (eds.), *Computational Statistics and Data Intelligence*, Springer Proceedings in Mathematics & Statistics 463,
https://doi.org/10.1007/978-981-97-4438-1_13

List of Symbols

b	Hertzian contact half-width (m)
E_t, E_i, E_k	Specific-total-energy, specific-internal-energy, specific-kinetic-energy (m^2s^{-2})
$\mathbf{f}^B$, $\mathbf{f}^S$	Vector of volume force in the fluid and solid (N m^{-3})
Fr	Froude number, $Fr = u_m/\sqrt{gh}$
h_{cen}-C	CFT in the constant VPC condition (μm)
h_{cen}-V	CFT in the virable VPC condition (μm)
Re	Reynolds number, $Re = (\rho u_m h)/\mu$
S_0	Viscosity-temperature index, $S_0 = \beta(T_0 - 138)/(\ln \eta_0 + 9.67)$
$\mathbf{V}$	Tensor of velocity (m/s)Tensor of velocity (m/s)
Z_0	Viscosity-pressure index, $Z_0 = a/P_0(\ln \eta_0 + 9.67)$
$\boldsymbol{\xi}$	Strain rate tensor (s^{-1}), $\xi = (\nabla\mathbf{V} + \nabla\mathbf{V}^T)/2$
$\dot{\gamma}_{eq}$	Shear strain rate tensor
η_N	Newtonian fluid viscosity (Pa s)
λ	Second coefficients of fluid viscosity
μ	Coefficients of fluid viscosity
$\underline{\tau}_f$, $\underline{\tau}_s$	Fluid and solid stresses on the fluid solid coupling surface (Pa)

1 Instruction

In 1966, Roelands pressure-viscosity correlation [1], an empirical equation, was proposed by Roelands and further developed by Houpert [2] to describe the viscosity variation of mineral oils with pressure and temperature. This equation is generally considered to have higher accuracy than the exponential equation [3], and has been widely used in the solution of EHL problems [4].

Viscosity-pressure coefficient (VPC) is a very important parameter in Roelands equation and used to be assumed as a constant, which however in fact varies with pressure and temperature. Roelands [1] clearly pointed out in his doctoral dissertation (1966) that the VPC is very small, and can be regarded as a constant in a wide extent of temperature and pressure (temperature ranges from 20 to 120°C, pressure ranges from 0 to 196 MPa) without significantly affecting the accuracy of the viscosity pressure temperature relationship. However, the EHL operation conditions of higher pairs in gears and bearings are generally beyond that limits, especially for the pressure [5, 6]. Consequently, this paper studies the effectiveness of Roelands formula with constant VPC in EHL calculation of higher Pair.

2 Modeling Approach

As shown in Fig. 1, the physical model of EHL point contact consists of elastic ball Ω_1, elastic plate Ω_2 and lubricating oil Ω_3. The ball rotates around the Z-axis which passing through its center O and normal to X–Y plane. The velocity of the ball is $u_1 = \omega R$, with R and ω the ball radius and the angular velocity. The plate moves in X-direction with the velocity u_2. The entrainment velocity and slide-to-roll ratio are defined as $u_m = (u_1 + u_2)/2$ and $\text{SRR} = (u_1 - u_2)/u_m$. Load F acts vertically on the surface of the ball.

A localized EHL point contact model was created, encompassing the contact region of the global model, as indicated by the red dashed box in Fig. 1. The creation of the local model is discussed in Ref. [7]. This model consists of both a structural module (comprising local ball and local plate models) and a fluid module (comprising a local lubricant model), as illustrated in Fig. 2. The local model has the dimensions of L = 400 μm in X-direction, as well as Z-direction, to ensure that distances larger than 6b [8]. Additionally, the thicknesses of the local plate and ball model in Y-direction must exceed 3.15b to ensure there is no temperature gradient ($\partial T/\partial Y = 0$) exists at the model boundary [9, 10]. Simultaneously, a depth of D = 400 μm is chosen to provide sufficient space for deformation.

Within the fluid module, the upper and lower surfaces function as interfaces facilitating fluid–structure interaction. These interfaces are designed as thermal shells, replicating the transfer of energy produced by fluid shear effects through the coupling interface via a virtual wall of a predetermined thickness. Two surfaces perpendicular to Z axis are set to uniform-flow, and the rotation and the movement of the part ball and disc are achieved by using the Non-slip movement on the fluid–solid-interaction

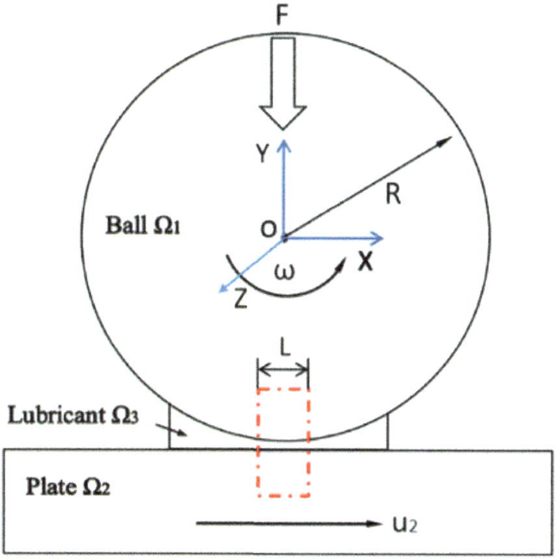

Fig. 1 Physical model of EHL point contact

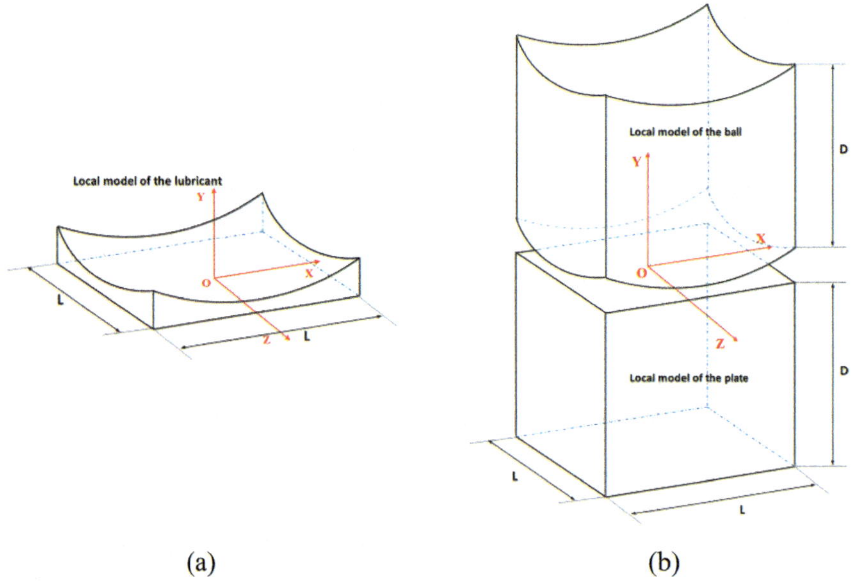

Fig. 2 Local model of point contact: **a** fluid model and **b** structural model [5]

surface. The surfaces perpendicular to the direction of the entrainment speed (X-axis direction) are set as the inlet and outlet, on which Dirichlet boundary conditions ($P_{in} = P_{out} = P_0$, $T_{in} = T_{out} = T_0$) are applied. The initial temperature of the whole calculation domain of the fluid module is set to the ambient temperature ($T_{oil} = T_0$). Assume that the contact area of the local model is fully lubricated.

Within the structural module, we designate the points of contact between the fluid module and the structure as the fluid–structure-interaction surfaces. Specifically, the bottom of the plate part model has fixed nodes. The freedom for all other nodes in both the local ball and disc model are freed up along the Y-axis, while restricting the remaining DOFs.

3 Fluid–Structure Coupling Governing Equations

Fluid Governing Equation

Fluid governing equation mainly refers to the three conservation functions in hydromechanics (continuity equation, momentum equation and energy equation), rheological equation and density equation.

The continuity equation can be deduced based on the principle that the net mass flow rate exiting the fluid element equals the reduction in its mass:

$$\frac{\partial \rho}{\partial t} + \nabla \cdot (\rho \mathbf{V}) = 0 \tag{1}$$

where ρ is fluid density, t is time, $\mathbf{V}$ is the fluid velocity tensor.

The principle of momentum conservation can be grasped as the concept the force acting on a fluid equals the momentum change rate.

$$\frac{\partial \rho \mathbf{V}}{\partial t} + \nabla \cdot (\rho \mathbf{V} \otimes \mathbf{V} - \boldsymbol{\tau}) = \mathbf{f}^{\mathbf{B}} \tag{2}$$

where $\mathbf{f}^{\mathbf{B}}$ is the vector of body force, $\boldsymbol{\tau}$ is the fluid stress tensor and is defined as $\boldsymbol{\tau} = (-P + \lambda \nabla \cdot \mathbf{V}) + 2\mu\boldsymbol{\xi}$, μ, λ and $\boldsymbol{\xi}$ are respectively molecular viscosity coefficient, second molecular viscosity coefficient and velocity strain tensor ($\boldsymbol{\xi} = (\nabla\mathbf{V} + \nabla\mathbf{V}^T)/2$).

The energy content per unit mass of a fluid element comprises internal energy and kinetic energy, which can be altered through processes such as heat generation and conduction, and the work performed by the resulting force. The fluid's inertia is taken into account, while the influence of gravity is omitted in the model based on the Froude number analysis ($Fr = u_m/\sqrt{gh_{eq}} >> 1$). The lubricant is considered as laminar flow for its model Reynolds number ($\mathrm{Re} = (\rho u_m h_{eq})/u < 1$) is no more than 50 in this work.

$$\frac{\partial \rho E_t}{\partial t} + \nabla \cdot (\rho \mathbf{V} E_t - \boldsymbol{\tau} \cdot \mathbf{V} + \mathbf{q}) = \mathbf{f}^{\mathbf{B}} \cdot \mathbf{V} + q^B \tag{3}$$

where ρ, u_m are density and entrainment velocity of the lubricating oil, $u_m = (u_1 + u_2)/2$, h_{eq} is the characteristic length of the fluid model, μ is lubricant viscosity coefficient, g is the gravitational acceleration, $\mathbf{q}$ is defined as heat flux, which assumed to conform to the Fourier's law: $\mathbf{q} = -k\nabla T$, where T is temperature, k is thermal conductivity. The specific total energy $E_t = \frac{1}{2}\mathbf{V} \cdot \mathbf{V} + E_i = E_k + E_i$. q^B is heat generation rate, defined as $q^B = Q_{shear} + Q_{compr}$, Q_{shear} and Q_{compr} represent the shear and compression heating rate. In this work, only Q_{shear} was considered according to Refs. [11, 12], and can be calculated through Formula (4) [8].

$$Q_{shear} = \mu\left(\nabla\mathbf{V} \otimes \nabla\mathbf{V} + \nabla\mathbf{V} \otimes (\nabla\mathbf{V})^T - \frac{2}{3}(\nabla\mathbf{V})^2\right) \tag{4}$$

where μ is fluid dynamic viscosity.

The relationship delineating the dependence of lubricant viscosity on pressure and temperature is formally characterized by Roelands' equation.

$$\eta(p, T) = \eta_0 \exp\left\{(\ln \eta_0 + 9.67) * \left[\left(1 + 5.1 * 10^{-9} * P\right)^{z_0} * \left(\frac{T - 138}{T_0 - 138}\right)^{-s_0} - 1\right]\right\} \tag{5}$$

where T_0, η_0 are reference temperature and viscosity; Z_0, S_0 reflect the relationships between viscosity and pressure, temperature, and can be respectively obtained by $Z_0 = \alpha C_V/5.1E - 09$ and $S_0 = \beta(T_0 - 138)/C_v$, with α VPC, β viscosity-temperature coefficient (VTC) and $C_v = \ln \eta_0 + 9.67$.

Table 1 displays physical parameters of lubricating oil under different temperatures, and the VPC is changing with temperature. The quadratic equation can be obtained by fitting the VPCs at different temperatures with the Adj R-Square of 0.99776.

$$a = 0.00406T^2 - 1.14937T + 183.76513 \quad (6)$$

The Ree-Eyring sinh law: $\eta(p, T, \dot{\gamma}_{eq}) = (\tau_0/\dot{\gamma}_{eq}) \sinh^{-1}(\eta_N \dot{\gamma}_{eq}/\tau_0)$ is employed to consider the non-Newtonian effects in the lubricants, where τ_0 is Ree–Eyring stress, $\dot{\gamma}_{eq} = (\tau_0/\eta_N) \sinh(\tau_{eq}/\tau_0)$ is shear strain rate tensor.

The changing of lubricant density with pressure and temperature is calculated by:

$$\rho(p, T) = \rho_0\left[1 + \frac{AP}{1 + BP} + D(T - T_0)\right] \quad (7)$$

where ρ_0 is reference density, A = 0.6E-09 1/Pa, B = 1.7E-09 1/Pa, D = –0.00065 1/K.

Table 1 Physical parameters of lubricating oil according to temperature

Temperature T(K)	Density ρ(Kg/m^3)	Dynamic viscosity μ(Pa s $\cdot 10^3$)	Thermal conductivity k(N/s°C)	VPC a(1/Pa $\cdot 10^{10}$)	VTC β(1/K)
283	1001.20	94.58	0.1672	174.48	4105
293	993.70	58.30	0.1650	161.79	3918
303	986.20	37.87	0.1627	151.25	3743
313	978.74	25.52	0.1605	142.56	3579
323	971.24	18.03	0.1583	134.86	3425
333	963.70	13.33	0.1561	128.50	3281
343	956.20	10.06	0.1539	122.98	3145
353	948.70	7.80	0.1517	118.19	3017
363	941.20	6.19	0.1494	114.01	2896
373	933.70	5.02	0.1472	110.34	2781
383	926.20	4.15	0.1450	107.10	2673
393	918.70	3.49	0.1428	104.23	2569

3.1 Solid Governing Equation

The solid constituents undergo deformation and spatial displacement in response to the exerted fluid pressure, with their movements adhering to the principles articulated in Newton's second law [13].

$$\rho_s \ddot{\mathbf{d}} = \nabla(\sigma_s) + \mathbf{f}^{\mathbf{S}} \tag{8}$$

where ρ_s, $\mathbf{f}^{\mathbf{S}}$ is density and external volume force vector of solid, σ_s is Cauchy stress tensor, $\ddot{\mathbf{d}}$ is solid gravity acceleration.

3.2 Fluid–Structure Coupling Equation

The kinematic condition and dynamic condition need to be satisfied on the fluid–solid-interaction surfaces. The fluid velocity is determined by the kinematic condition when no-slip wall is used, and the fluid stress is integrated and exerted on the structure nodes in accordance with the dynamic condition.

$$\underline{\mathbf{d}}_f = \underline{\mathbf{d}}_s \tag{9}$$

$$\mathbf{n} \cdot \underline{\boldsymbol{\tau}}_f = \mathbf{n} \cdot \underline{\boldsymbol{\tau}}_s \tag{10}$$

$$\underline{\mathbf{V}} = \dot{\underline{\mathbf{d}}}_s \tag{11}$$

$$\underline{\mathbf{F}}(t) = \int h^d \underline{\boldsymbol{\tau}}_f \cdot dS. \tag{12}$$

where $\boldsymbol{d}$ and $\boldsymbol{\tau}$ are the displacement and stress, subscript s, f represent the fluid, solid respectively, $\underline{\mathbf{V}}$ is fluid velocity, h^d is virtual displacement of structural points, S is area of fluid, $\underline{\mathbf{F}}(t)$ is the fluid force acting on the structure.

4 Calculation Process

Figure 3 illustrates the procedural steps for calculating the EHL point contact. Initially, the geometric model is created and meshed, and the computational domain boundaries are defined using ADINA. Subsequently, properties such as generation of the heat, fluid characteristics, etc. are configured within the ADFUSR, which will be referenced in the subsequent calculations.

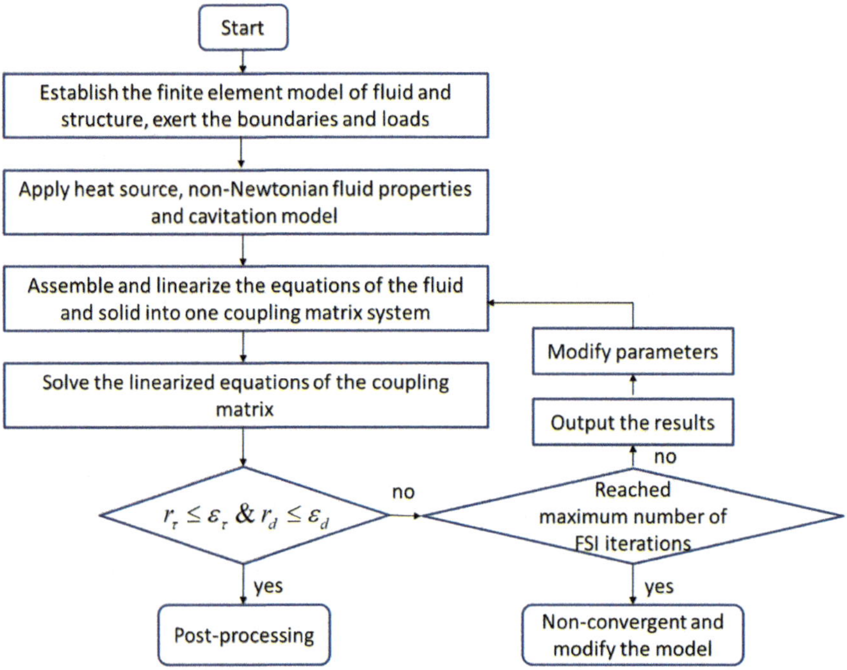

Fig. 3 Solution flow chart

5 Effectiveness of Roelands Formula with Constant VPC

According to the Roelands formula, the three-dimensional color map surface of lubricant viscosity with pressure and temperature when the VPC is 155E-10pa^{-1} (298 K) and 114E-10pa^{-1} (363 K) are presented in Fig. 4. The pressure range is 0–1000 MPa, and the temperature range is 298–398 K (25–125 °C).

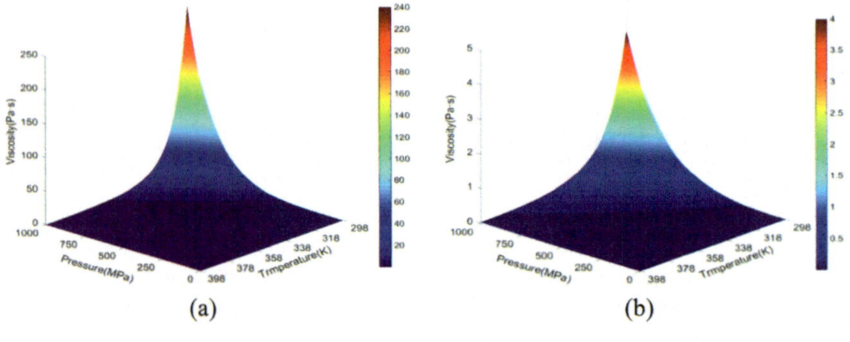

Fig. 4 Lubricant viscosity with **a** a = 155E-10pa^{-1}and **b** a = 114E-10pa^{-1}

When the pressure is lower than 400 MPa, the viscosities of lubricating oil under two different VPC conditions are very small throughout the whole temperature range. The influence of VPC on the viscosity of lubricating oil can be basically ignored, which is consistent with the conclusion of Roelands [1]. When the pressure is >400 MPa as well as the temperature is lower than 350 K, the viscosity of lubricating oil increases significantly with the increasing of pressure. Meanwhile, the lubricant viscosities of the two VPCs are differ greatly, and the difference becomes more obvious with the decline of temperature and the increase of pressure. Therefore, the accuracy of the lubricant viscosity will be seriously affected if the VPC is considered as a constant.

EHL point contact models with VPCs of 82E-10 pa^{-1}, 112E-10 pa^{-1}, 142E-10 pa^{-1} and 172E-10 pa^{-1} were solved to investigate the influence of the VPC on lubrication characteristic parameters. The SRR varies from 0 to 100%, $u_m = 1$ m/s, $F = 50$N, $R = 11.1125$ mm, $T_0 = 298$ K, $\eta_0 = 0.02552$ Pa · s. Take the VPC = 112E-10pa^{-1} for example, 3D film thicknesses and temperatures with SRR of 0% and 100% are separately shown in Figs. 5 and 6.

Contour maps depicting film thickness, employing two distinct SRRs, exhibit the traditional horseshoe pattern within the contact region, along with a notable

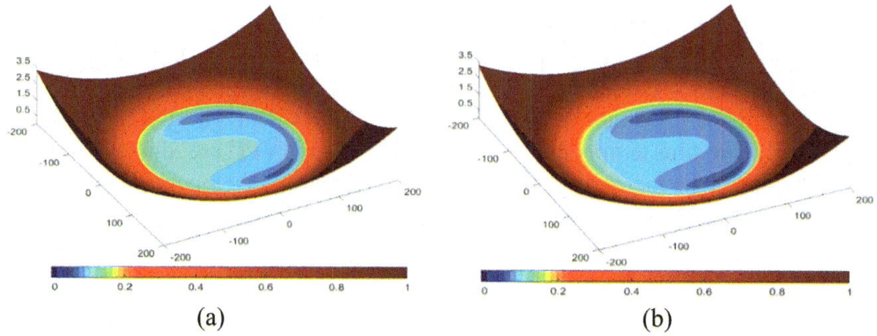

Fig. 5 3D film thickness contour maps with **a** SRR = 0% and **b** SRR = 100%

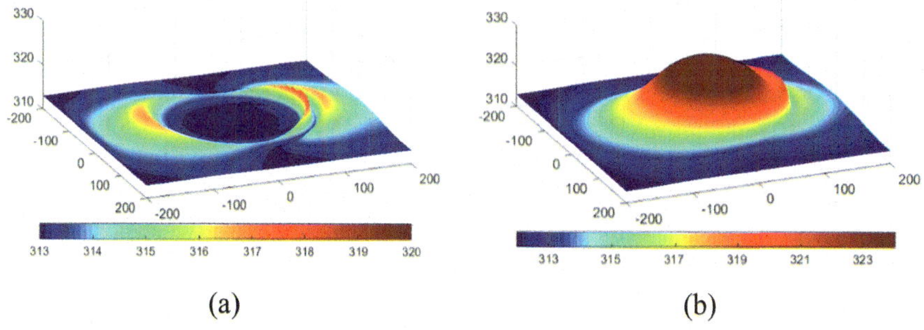

Fig. 6 3D film temperature contour map with **a** SRR of 0% and **b** 100%

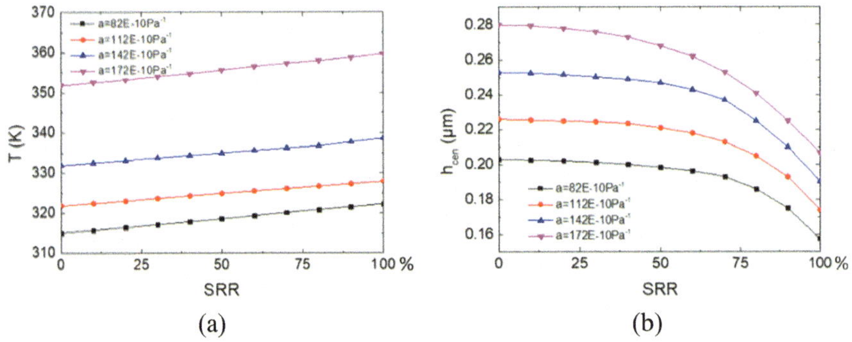

Fig. 7 **a** Film temperature and **b** CFT under different VPC

occurrence of necking at the termination of the contact area. The color of the contour map with SRR = 0% is a little darker compared to that of the SRR = 100% condition in the same scale, indicating that film thickness of the former is little greater than the latter. With the increase of SRR, the temperature rise of oil film increases. A secondary temperature spike, like the second pressure spike, appears near the necking area in the film temperature contour map [14].

CFTs and film temperatures under different SRRs are shown in the Fig. 7. Film temperature is highly sensitive to the variation of VPC. Even a modest increase in the VPC produces an important rise in film temperature in the higher pair. With the same VPC, the temperature of lubricating oil increases with the SRR basically at a fixed slope. The film temperature curves with different VPC are almost parallel to each other. With the increasing of the VPC, the sensitivity of temperature to the VPC increases, that is, the same increment of VPC will lead to greater temperature rise.

The CFT increases approximately linearly with VPC. The effects of the SRR on CFT are relatively insignificant until SRR reaches around 0.5. With the SRR increases, the viscous shear heating gradually becomes remarkable and the CFT presents shear-thinning behavior, resulting in the reduction of the CFT difference among the four VPCs. Consequently, considering the VPC as a constant in Roelands equation may lead to significant impacts on the film thickness and temperature rise in EHL solution of higher pair.

Comparisons of MFT and CFT between the constant VPC and variable VPC conditions were carried out, as shown in Fig. 8. The operating condition is u_m = 1 m/s, SRR = 100%, F = 20 N and the reference temperature changes from 283 to 333 K (corresponding VPC decreases from 174.48 E-10pa-1 to 128.5 E-10pa-1). At the same reference temperature, the CFT under the constant VPC condition is larger than that of variable VPC condition. As the reference temperature rises, the CFT decreases while the MFT increase grows. The gap in CFT between constant and variable VPC conditions gradually narrows, along with the reduction in maximum temperature rise.

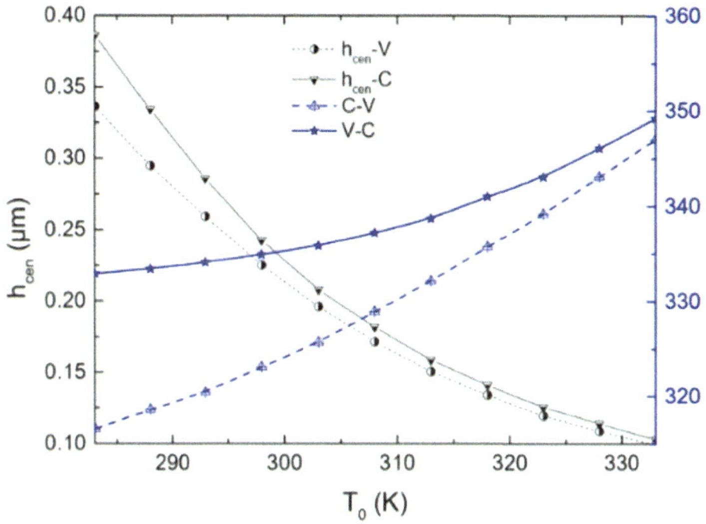

Fig. 8 Comparisons of MFT and CFT

6 Conclusion

The effectiveness of Roelands formula with a constant VPC in EHL solution of higher pairs is studied with different VPCs and SRRs, and conclusions are as follows:

(1) 1) The lubricant viscosity, CFT and MFT in the EHL point contact are highly sensitive to the VPC at low temperature and high pressure. Even a modest change in VPC produces an important rise/reduction of the EHL characteristic parameters.

(2) 2) When the reference temperature is small, the Roelands formula with constant VPC produces a certain deviation to the CFT and MFT in the higher pair. With the increase of reference temperature, the deviation decreases gradually.

Acknowledgements This work is supported by the National Key R&D Program of China (Grant No. 2022YFB4201100) and Chongqing NSF project (Grant No. CSTB2024NSCQ-BSX0002).

References

1. Roelands, C.J.A.: Correlational aspects of the viscosity-temperature-pressure relationship of lubricating oils. Delft University of Technology (1966)
2. Houpert, L.: New results of traction force calculations in EHD contacts. ASME J. Tribol. **107**, 241–248 (1985)
3. Olver, A.V., Spikes, H.A.: Prediction of traction in elastohydrodynamic lubrication. Proc. Inst. Mech. Eng. Part J—J. Eng. Tribol. **212**(5), 321–332 (1998)

4. Dowson, D., Ehret, P.: Past, present and future studies in elastohydrodynamics. Proc. Inst. Mech. Eng. Part J—J. Eng. Tribol. **213**(5), 317–333 (1999)
5. Huang, X., Yang, B., Wang, Y.: A nano-lubrication solution for high-speed heavy-loaded spur gears and stiffness modelling. Appl. Math. Model. **72**, 623–649 (2019)
6. Shi, X., Lu, X., He, T., et al.: Predictions of friction and flash temperature in marine gears based on a 3D line contact mixed lubrication model considering measured surface roughness. J. Central South Univ. **28**, 1570–1583 (2021)
7. Zhang, J., Dong, M., Hu, Y., et al.: Influence of coating material on elastohydrodynamic lubrication between gear meshing interfaces based on direct coupling solution. Adv. Comp. Hybrid Mater. **5**, 3117–3136 (2022)
8. Bruyere, V., Fillot, N., Morales-Espejel, G.E., Vergne, P.: Computational fluid dynamic and full elasticity model for sliding line thermal elastohydrodynamic contacts. Tribol. Int. **46**(1), 3–13 (2012)
9. Kaneta, M., Shigeta, T., Yang, P.: Film pressure distributions in point contacts predicted by thermal EHL analyses. Tribol. Int. **39**(8), 812–819 (2006)
10. Wang, Y., Li, H., Tong, J., Yang, P.: Transient thermoelastohydrodynamic lubrication analysis of an involute spur gear. Tribol. Int. **37**(10), 773–782 (2004)
11. Spikes, H.A., Anghel, V., Glovnea, R.: Measurement of the rheology of lubricant films within elastohydrodynamic contacts. Tribol. Lett. **17**(3), 593–605 (2004)
12. Reddyhoff, T., Spikes, H.A., Olver, A.V.: Compression heating and cooling in elastohydrodynamic Contacts. Tribol. Lett. **36**(1), 69–80 (2009)
13. Bathe, K.J.: Computational Fluid and Solid Mechanics, 2nd edn. Springer, Berlin (2011)
14. Zhang, J.: Elastohydrodynamic lubrication of meshing interface for high-speed and heavy-load gear based on finite element method. Chongqing University (2022)

Numerical Simulation of Oxygen Jet Condensation in Cryogenic Liquid Rocket

Chengfeng Zhu and Yanzhong Li

Abstract Low frequency pressure fluctuation is an important topic on the stable and safe working of the system. As the excitation source of the pressure fluctuation, oxygen injection condensation is the indispensable process in the pipeline system of liquid fuel aircraft. Due to the intense mass transfer and flow instability, it is difficult to capture the accurate pressure characteristic of oxygen jet condensation, especially in the first main frequency of pressure oscillation. Aiming at the mechanism and excitation process of pressure oscillation, a numerical simulation is carried out using a modified mass transfer model. The height function method is implemented to calculate the curvature of gas–liquid interface. It can update the mass transfer rate in real-time, which is the core technology to evaluate the pressure fluctuation. The modified model is verified by a water steam jet condensation simulation and the numerical result agrees well with the experimental data. The low-frequency pressure fluctuation characteristic is obtained successfully. The first main frequency is 9.5 Hz with the apparent amplitude of approximately 80 kPa. The research shows that the periodic mass transfer rate and the swing of the continuous oxygen gas plume are the key factor causing the low frequency pressure oscillation. There is an oxygen suck-back flow phenomenon in the oxygen chamber of condenser pipe. In addition, it is also found that decreasing the injection area can restrain the pressure oscillation effectively. These conclusions provide a theoretical guidance for overcoming the low-frequency pressure oscillation eventually.

Keywords Oxygen injection condensation · Height function · Heat and mass transfer · Low frequency pressure oscillation · Curvature estimation

C. Zhu (✉) · Y. Li
Xi'an Jiaotong University, Xi'an, China
e-mail: zhuchengfeng@stu.xjtu.edu.cn

W. Wang et al. (eds.), *Computational Statistics and Data Intelligence*, Springer Proceedings in Mathematics & Statistics 463,
https://doi.org/10.1007/978-981-97-4438-1_14

1 Introduction

Low-frequency pressure fluctuation in a liquid rocket is an unstable phenomenon caused by the dynamic interaction between the propulsion system and structural system [1]. In severe cases, it will endanger the safety of the rocket by arousing the self-excited pogo vibration. Its mechanism is not very clear yet. There is a general consensus [2] that the O2/LOX (oxygen and liquid oxygen) direct contact jet condensation in the propulsion pipe, as an excitation source, causes the low-frequency pressure fluctuation. It is important to explore the oxygen injection condensation process for revealing the mechanism of low-frequency pressure fluctuation phenomenon.

The oxygen injection condensation, as a kind of heat and mass transfer process, occurs with violent phase change and flow unsteadiness [3]. Recent researchers have mainly conducted different experiments, and few researchers focus on the numerical method about the pressure fluctuation in the oxygen jet condensation. Because it is difficulty to capture the pressure oscillation characteristic by the present mass transfer model [4]. For example, Li et al. [5] studied the contact condensation of water steam by the energy balance model and obtained an overestimated pressure oscillation frequency, 6 and 13 Hz. However, the experiment date of the frequency was 3.7 Hz and 9 Hz respectively, with a deviation about 44 ~ 62%. It is necessary to study the accurate algorithm to capture the pressure characteristic of contact condensation. Recently, Zhu et al. [6] proposed a modified mass transfer model based on the height function method. It is found that the new model can obtain a more accurate result of pressure oscillation than other numerical models. Meanwhile, the model is verified by the oxygen bubble condensation simulation [7] with the excellent agreement.

In this paper, the Zhu's mass transfer model is used for the oxygen contact condensation simulation to investigate the mechanism and influence factors of low frequency pressure oscillation. The research results provide guidance for overcoming the low-frequency pressure fluctuation in the propulsion system completely.

2 Calculation Procedure

2.1 Numerical Method

This paragraph elaborates the numerical model for mass transfer rate simulation and the detailed numerical procedure can be referred to Zhu's model [6]. This model can provide more accurate pressure values than conventional models [6]. The height function method is implemented to calculate the curvature of two-phase interface. It is the key to capture the precise low frequency pressure fluctuation in this numerical model. The interfacial curvature κ can be deduced as follow,

$$\kappa = -\nabla \cdot \hat{\mathbf{n}} = \frac{H_{xx}(1 + H_y^2) + H_{yy}(1 + H_x^2) - H_x H_y (H_{xy} + H_{yx})}{(1 + H_x^2 + H_y^2)^{3/2}} \quad (1)$$

where the term H is the variable of fluid height in the height function method, $\hat{\boldsymbol{n}}$ is the unit normal vector of gas–liquid interface. Then the specific volumetric mass source term S_m and heat source term S_h is shown in the follow form,

$$S_m = 3\kappa k\alpha(\nabla T \cdot \hat{\mathbf{n}})/h_L \tag{2}$$

$$S_h = S_m h_L \tag{3}$$

where the h_L represents the heat of latent, k is the heat conduction rate, and α is the oxygen volume fraction. The volume of fluid methodology is used to calculate the gas–liquid interface. The mass and heat source term are then implemented into the continuity, momentum and energy equations, which are shown as follow.

$$\frac{\partial}{\partial t}(\alpha\rho) + \nabla \cdot \left(\alpha\rho\vec{v}\right) = S_m \tag{4}$$

$$\frac{\partial}{\partial t}\left(\rho\vec{v}\right) + \nabla \cdot \left(\rho\vec{v}\vec{v}\right) = -\nabla p + \nabla \cdot \left[\mu\left(\nabla\vec{v} + \nabla v^T\right)\right] + \rho\vec{g} + \vec{F}_{vol} \tag{5}$$

$$\frac{\partial}{\partial t}(\rho E) + \nabla \cdot \left[\vec{v}(\rho E + p)\right] = \nabla \cdot \left(k_{eff}\nabla T\right) + S_h \tag{6}$$

where ρ is the density of fluid, μ is the dynamic viscosity, k_{eff} is the effective thermal conductivity. The $\vec{F}_{vol}$ represents the surface tension across the gas–liquid interface and it is obtained from the continuum surface force model. The ρ_m is the average density of fluid, ρ_v is the density of vapor, ρ_l is the density of liquid.

$$\vec{F}_{vol} = \sigma\frac{2\rho_m\kappa\nabla\alpha}{\rho_v + \rho_l} \tag{7}$$

2.2 Physical Model

The oxygen injection condensation is always occurred in the condenser pipe in the propulsion system. Figure 1 exhibits a sketch diagram of the specific propulsion system and the corresponding structure of the condenser pipe. The condenser pipe is installed between the two pumps, highlighted in black dotted line. The condenser pipe contains an annular gas and liquid entrance and a circular exit. The other boundary is regarded as wall. The mass flow rate is set for the inlet and the pressure vent is set for the liquid outlet. The oxygen flows from the annular gas entrance into the bend chamber across the two rows of small holes. Then the entire oxygen is condensed into liquid oxygen. There are 32 holes in each row distributed on the wall of bend

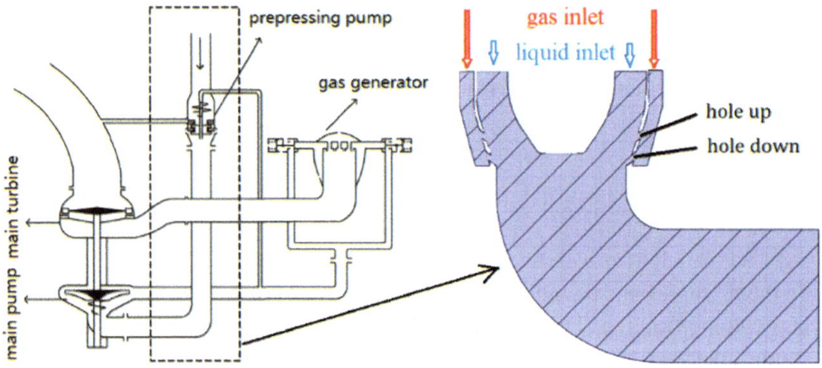

Fig. 1 A sketch view of propulsion system and condenser pipe (in the dotted box)

chamber. There are two monitor points placed in the center of two rows of injection hole, named "hole up" and "hole down" respectively.

Table 1 exhibits the specific working conditions. The liquid oxygen has a subcooling of about 33 K. The NIST PROP database is used for the physical properties of oxygen and liquid oxygen. There are five cases arranged for the comparison and analyze. The case A is the control group. In case A and B, the oxygen status, liquid oxygen status and hole diameter are the same. The difference is the hole number, which cause the difference of oxygen mass flux.

Table 1 The detailed operation conditions in different cases

Case item	Oxygen status	Liquid oxygen status	Hole numbers	Hole diameter	Oxygen mass flux
Case A	1.5 kg·s^{-1}, 150 K	90 kg·s^{-1}, 90 K	32 × 2	8 mm	85 kg·m^{-2}·s^{-1}
Case B	1.5 kg·s^{-1}, 150 K	90 kg·s^{-1}, 90 K	16 × 2	8 mm	170 kg·m^{-2}·s^{-1}
Case C	0.75 kg·s^{-1}, 150 K	90 kg·s^{-1}, 90 K	16 × 2	8 mm	85 kg·m^{-2}·s^{-1}
Case D	1.5 kg·s^{-1}, 150 K	90 kg·s^{-1}, 90 K	32 × 1	8 mm	170 kg·m^{-2}·s^{-1}
Case E	0.75 kg·s^{-1}, 150 K	90 kg·s^{-1}, 90 K	32 × 1	8 mm	85 kg·m^{-2}·s^{-1}

3 Result and Discussion

In order to investigate the oxygen jet condensation process visually, Fig. 2 illustrates the distribution of oxygen volume fraction for case A, B and C. Case A and B keeps the constant oxygen mass flow rate, approximately 1.5 kg·s^{-1}, while case B contains smaller hole numbers with a higher oxygen mass flux. The length of oxygen plume of case B is longer than that of case A.

Figure 3 illustrates the obvious pressure oscillation characteristic obtained by the numerical model in different cases. It is clear that there is intent pressure oscillation with an apparent amplitude of approximate 80 kPa in case A, while the pressure is steady in case B and C without obvious fluctuation. The pressure in case B and C keep the back pressure 1220 kPa. The apparent amplitude of pressure oscillation in case B and C is only 0.6 kPa. This is because the hole number in case B and C is a half of that in case A. The mass flux of case B is larger than that of case A. The mass flow rate of C is smaller than that of case A. These factors cause the disappearance of pressure oscillation in case B and C, compared with the case A. Only in case A, the pressure is unsteady with the flow instability.

(a) case A (b) case B (c) case C

Fig. 2 Distribution of oxygen volume fraction in different cases

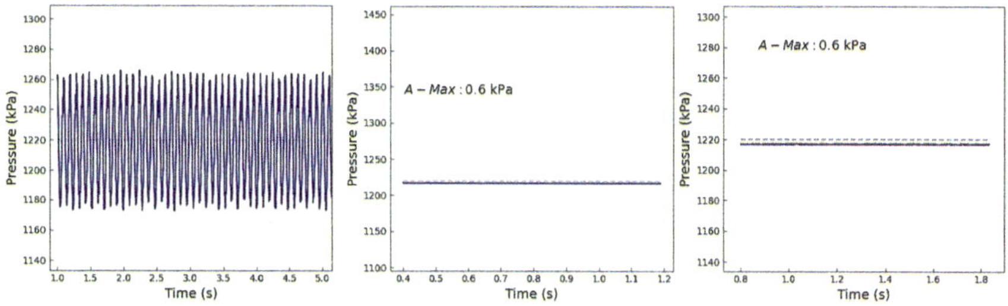

(a) case A (b) case B (c) case C

Fig. 3 The pressure oscillation varying with time in different cases

In order to explore the pattern of different pressure fluctuation in case A, B and C, the corresponding oxygen velocity, density and temperature in monitor points are extracted in Fig. 4. In case A, the physical field shows strong instability, on account of the oxygen flow condensation in the condenser pipe. As for the temperature distribution, it is found that the temperature in monitor "hole up" fluctuates with time from 115 to 90 K periodically. When the temperature is 90 K, it means that the liquid oxygen is sucked back to the pipe. When the temperature is 115 K, the oxygen column crush into the main pipe again. In this way, the flow instability comes out.

Compared with case A, the physical field keeps constant with time in case B and C. The temperature in monitor points keeps approximately 150 K. It shows that the liquid oxygen suck-back phenomenon is not occurred in case B and C. This is because the hole numbers are small with a small injection area. The oxygen inject velocity is not large enough to arouse the period oscillation. The oxygen density is about 33 kg/m^3 in case B and C. The oxygen does not show its compressibility. The oxygen velocity in case B is 35 ~ 40 m/s, while that of case C is 18 ~ 20 m/s.

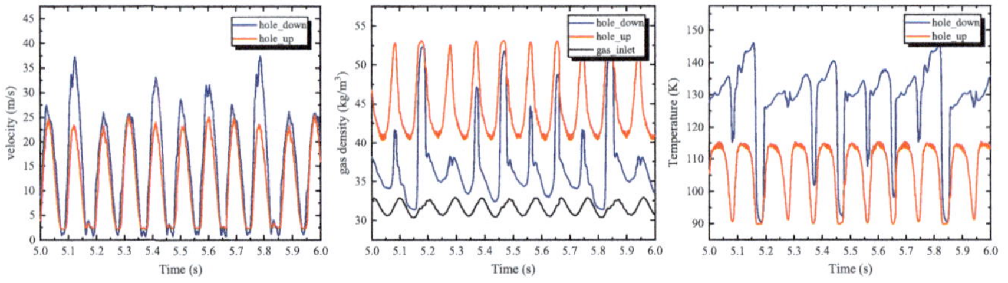

(a) case A

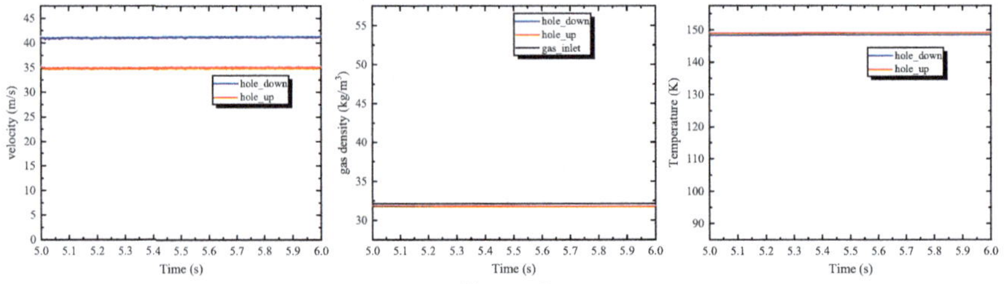

(b) case B

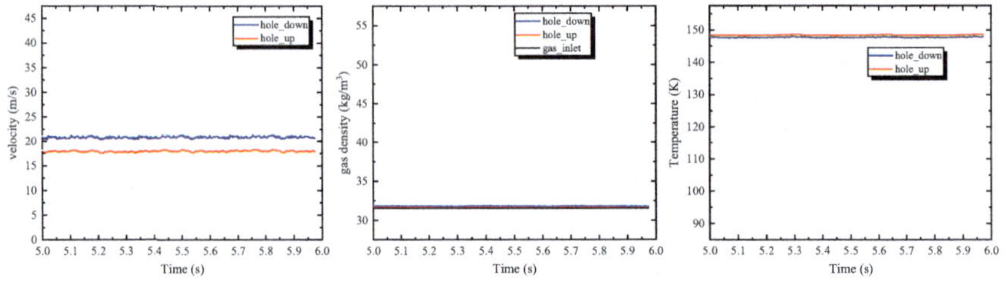

(c) case C

Fig. 4 The velocity, density and temperature in monitor points varying with time

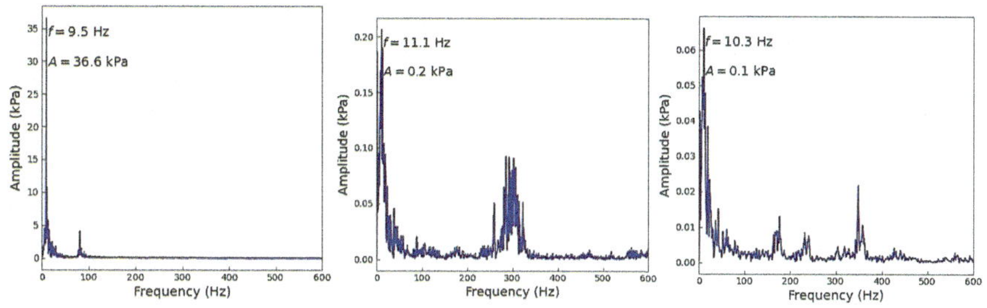

(a) case A (b) case D (c) case E

Fig. 5 Distribution of spectral analysis characteristic in different cases

By the spectral analysis, Fig. 5 illustrates the distribution of frequency and amplitude in different cases. In case A, the first main frequency is 9.5 Hz, with the main amplitude of 35 kPa. In case D and E, the first main frequency is 10 ~ 11 Hz, with the main amplitude of below 0.2 kPa. It also demonstrates that the oxygen jet condensation is stable in case D and E.

Same as case B and C, Fig. 6 illustrates the distribution of oxygen inject velocity, density and temperature varying with time in case D and E. In case D, the average velocity of oxygen is about 40 m/s, while that of case E is 18 m/s. This is because the oxygen mass flow rate in case D is 1.5 kg/s, while that in case E is 0.75 kg/s. In addition, the oxygen density keeps 30 kg/m^3 and the oxygen temperature keeps about 150 K in both case D and E. There is no obvious pressure oscillation in case D and E. This is because the two rows of holes have been changed into one row with smaller injection area in case D and E. When the injection area is smaller with the same oxygen mass flow rate, the oxygen mass flux will be larger, resulting in the fact that the oxygen velocity is faster. The flow will be steady without obvious fluctuation.

Above all, the injection area will influence the oxygen injection condensation by changing the oxygen mass flux. It is clear that decreasing the injection area can restrain the pressure oscillation under the same oxygen mass flow rate.

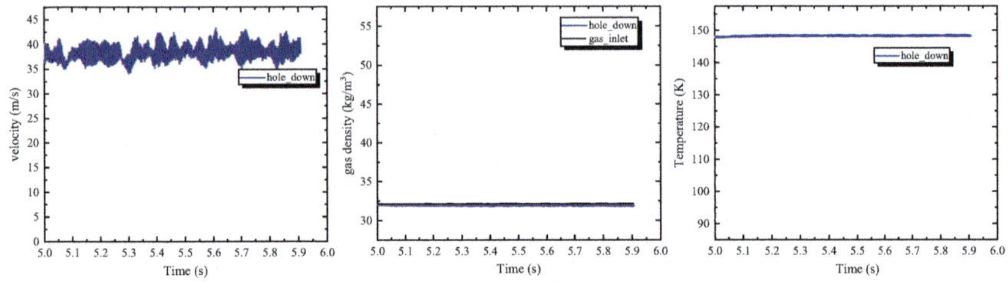

(a) case D

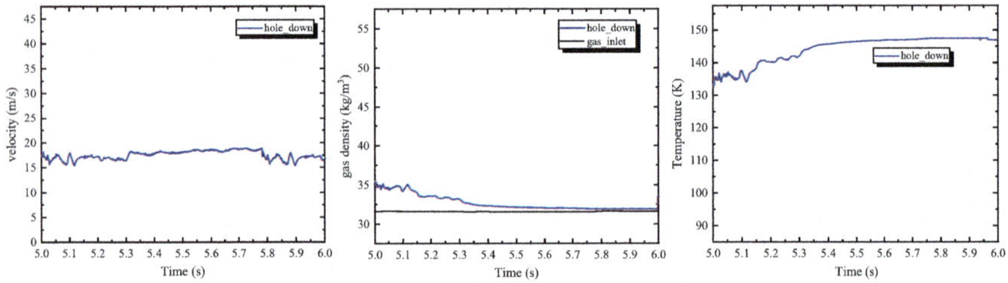

(b) case E

Fig. 6 The velocity, density and temperature varying with time in case D and E

4 Conclusion

In this work, a numerical simulation about the oxygen contact condensation is carried out by the modified energy balance mass transfer model. The low-frequency pressure fluctuation characteristic is captured accurately. The first main frequency is 9.5 Hz with the apparent amplitude of approximately 80 kPa. It is also found that decreasing the injection area can restrain the pressure oscillation effectively. The research results provide guidance for the structure design of propulsion system.

References

1. Kim, Y., Roh, T., Lee, H.: Mechanism and prediction of flow oscillation based on counter-flow jet for drag reduction in hypersonic flow. Aerosp. Sci. Technol. **126**, 107603 (2022)
2. Zhu, C., Li, Y., Xie, F., et al.: The mechanism research of low-frequency pressure oscillation in the feeding pipe of cryogenic rocket propulsion system. Processes **10**(11), 2448 (2022)
3. Chong, D., Liu, W., Zhao, Q., et al.: Oscillation characteristics of periodic condensation induced water hammer with steam discharged through a horizontal pipe. Int. J. Heat Mass Transf. 173 (2021)
4. Tanskanen, V., Jordan, A., Puustinen, M., et al.: CFD simulation and pattern recognition analysis of the chugging condensation regime. Ann. Nucl. Energy **66**, 133–143 (2014)
5. Li, S., Han, W., Lu, T., et al.: Numerical study on oscillation characteristics of large chugging of direct condensation of steam in a Tee junction with flowing sub-cooled water. Prog. Nucl. Energy 135 (2021)

6. Zhu, C., Li, Y., Xie, F., et al.: Development of a modified mass transfer model based on height function method to capture the pressure oscillation occurred in direct contact condensation process. Int. J. Heat Mass Transf. **201**, 123619 (2023)
7. Zhu, C., Li, Y., Xie, F., et al.: Numerical investigation of oxygen bubble condensation in subcooled quiescent liquid oxygen based on height function method. Appl. Therm. Eng. **225**(5), 120207 (2023)

Numerical Simulation of Mechanical Bulging Process of a Large Three-Layer Rectangular Bellow

Hong Jiang, De-an Meng, Kun Li, Chengcheng Zhu, Hao Zhou, and Shengdun Zhao

Abstract A novel mechanical bulging process is proposed in this work to overcome the disadvantages of the existing forming process for large rectangular bellows with deep waves. Numerical simulation was conducted on the mechanical bulging process of the three-layer 304 stainless steel rectangular bellow. Simulation results revealed that each layer steel had similar deformation and equivalent strain distribution. The maximum equivalent strain at the round corner of the rectangular bellow was 0.406. The general dimensions of the rectangular bellow were well consistent with the designed values. The wall thickness thinning rate at the round corner was greater than that in the straight edges, which had a maximum value of 20.48%.

Keywords Rectangular bellow · Stainless steel · Mechanical bulging · Equivalent strain · Wall thickness

1 Introduction

Metal bellows are elastic components used to relieve vibration and compensate displacement, widely used in machinery, chemistry, electric power, automotive industry, aerospace [1–3]. The quality and performance of the bellows are closely associated with the forming process. The most common process for fabricating metal bellows is hydroforming which can obtain good surface quality and accurate dimensions, and there are some other forming methods which combines magnetoelectricity and local heating. Jiang et al. [4] and Liu et al. [5] investigated the influence of process parameters on the wall thickness and the deformation behavior of metal during the hydroforming process of 304 stainless steel bellows. Hu et al. [6] studied the effect of magnetorheological elastomer (MRE) and axial feeding on the forming

H. Jiang · K. Li · H. Zhou · S. Zhao (✉)
Xi'an Jiaotong University, Xi'an, Shangxi 710049, China
e-mail: sdzhao@mail.xjtu.edu.cn

D. Meng · C. Zhu
Chang'an University, Xi'an, Shangxi 710064, China

W. Wang et al. (eds.), *Computational Statistics and Data Intelligence*, Springer Proceedings in Mathematics & Statistics 463,
https://doi.org/10.1007/978-981-97-4438-1_15

quality of thin-walled 718 Inconel bellows. Zhang et al. [7] proposed a buckling-induced forming method including local heating and cooling to fabricate 304 stainless steel bellows. Yang et al. [8] used a current-assisted forming technology to fabricate titanium alloy bellows.

Large rectangular bellows with deep waves are used in some special and important devices, such as International Thermonuclear Experimental Reactor (ITER), to absorb the relative displacements between the vacuum vessel and the cryostat resulting from thermal expansion [9, 10]. However, there are many shortcomings in these existing methods for fabricating such large rectangular bellows. Large pump stations and cylinders are required in hydroforming process, where hydraulic oil leakage has always been a highly focused concern. Furthermore, hydroforming process is more suitable for cylindrical bellows. MRE is suitable for bellows with small deformation and low wave crests due to the sealing and the ability of the MRE to transfer pressure. Uneven deformation and cracks may occur at the interface of heat and cold in the local heating and cooling method. Current-assisted forming technology may has the potential of uneven heating and changes the microstructure of the billet. These deficiencies are not conducive to the fabrication of large rectangular bellows, and an eco-friendly and accurate forming method is urgently needed. In this work, a novel mechanical bulging process is proposed, which can form multiple waves in turn by arranging concave and convex dies outside and inside of the rectangular billet. There is no hydraulic oil leakage problem and the waveforms are mainly controlled by the dies. This study presents a numerical simulation of the mechanical bulging process of a large three-layer stainless steel 304 rectangular bellow. The geometry, equivalent strain, and wall thickness of the rectangular bellow during the forming process were analysed in detail.

2 The Scheme and Finite Element Modeling

2.1 *The Scheme of Mechanical Bulging Process*

The scheme for the mechanical bulging process of the rectangular bellow is shown in Fig. 1, which shows the longitudinal section of the rectangular bellow and the movement of the dies. There are four sets of concave dies and three sets of convex dies located outside and inside the billet, and each wave is achieved by controlling the trajectories of these dies. In the initial state, the lower edge of the billet is fixed on the ground. In pass 1, the upper concave die 1 (UCCD1), middle concave die 2 (MCCD2), the upper convex die 1(UCVD1), and middle convex die 2 (MCVD2) maintain the initial positions, whilst the middle concave die 3 (MCCD3), lower concave die 4 (LCCD4)) and lower convex die 3 (LCVD3) are moved down along the vertical direction, and LCVD3 is also moved horizontally along the outside of the billet, to form the first wave. In pass 2, the UCCD1, MCCD3, LCCD4, UCVD1, and LCVD3 are maintained in the same positions of pass 1, whilst MCCD2 and

MCVD2 are moved down along the vertical direction, and MCVD2 is also moved horizontally along the outside of the billet. Similarly, in pass 3, only UCCD1 and UCVD1 are moved to the desired positions to form the third wave.

The geometry of the rectangular bellow is shown in Fig. 2. The overall dimension of the three-layer rectangular bellow in this work is 3435 mm × 2714 mm × 310 mm, the wall thickness of the rectangular bellow is 4.5 mm, and the number of the waves is 3. The initial billet is a rectangular billet with three layers of 304 stainless steel, and the thickness of each layer is 1.5 mm.

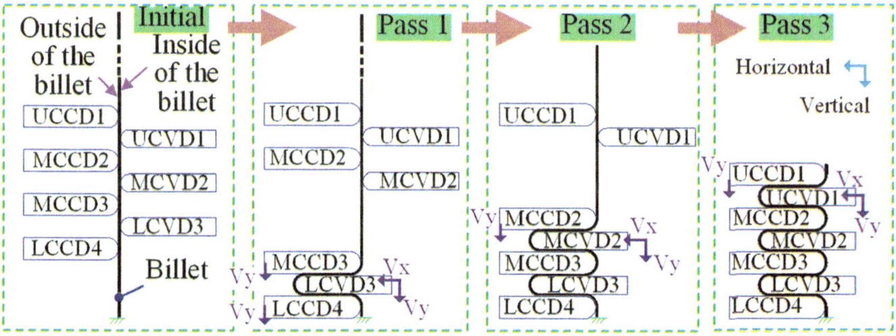

Fig. 1 Scheme for the mechanical bulging process of a three-layer rectangular bellow

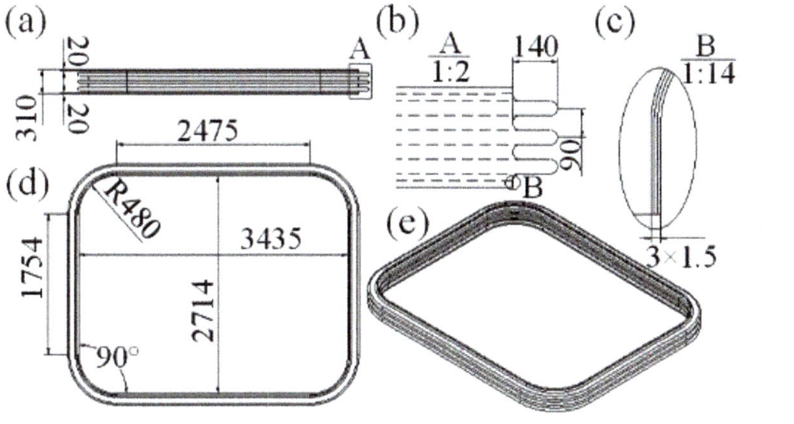

Fig. 2 The geometry of the rectangular bellow: **a** height view; **b** enlargement of region A; **c** enlargement of region B; **d** top view; **e** 3D view

2.2 The FE Model of the Mechanical Bulging Process

The simulation of the mechanical bulging process was numerically analysed by ABAQUS/Explicit. The quarter of the billet with symmetry boundary conditions was used to reduce the calculation time. There were several sets of concave and convex dies arranged on both sides of the billet, and flange edges were arranged on the upper and lower edges of the billet. The FE model is displayed in Fig. 3. 304 stainless steel was used as the material of the three-layer billet (E = 206 GPa, μ = 0.3). Johnson–Cook constitutive model was applied to the numerical simulation, as shown in Eq. (1).

$$\sigma_{eq} = \left(A + B\varepsilon_{eq}^{n}\right)\left[1 + Cln\left(\dot{\varepsilon}_{eq}/\dot{\varepsilon}_0\right)\right]\left[1 - \left(T^*\right)^m\right] \tag{1}$$

where σ_{eq}^{n} is the equivalent flow stress, A is the static yield stress, B is the strength coefficient, C is the strain rate sensitivity coefficient, m is the thermal sensitivity coefficient, n is the strain-hardening exponent, ε_{eq} is the equivalent plastic strain, $\dot{\varepsilon_{eq}}$ is the equivalent strain rate, $\dot{\varepsilon_0}$ is the reference strain rate, $T^* = (T - T_r)/(T_m - T)$, where T_r, T_m and T are the reference temperature, the melting temperature and the deformation temperature, respectively. The parameters of the J–C model were: A = 566 MPa, B = 832.8 MPa, n = 0.5647, C = 0.02876, and $\dot{\varepsilon_0} = 0.008s^{-1}$ [11].

The inner, middle and outer layer of the three-layer rectangular billet were regarded as layer 1–3, respectively. Shell elements (element size = 5 mm) and R3D4 elements were assigned for the billet and all dies, respectively. The total element number of the model is 421025. The contact type was 'surface to surface'. The frictions between the billet and dies and between each layer were Coulomb, with a coefficient of 0.1. The symmetrical boundaries along the XY and YX planes were set at the longitudinal edge of the quarter of the billet. The upper and lower flanges were tied with the inner layer, the former was followed with the billet whilst the latter was fixed on the ground. The movements of the dies were set according to Fig. 1.

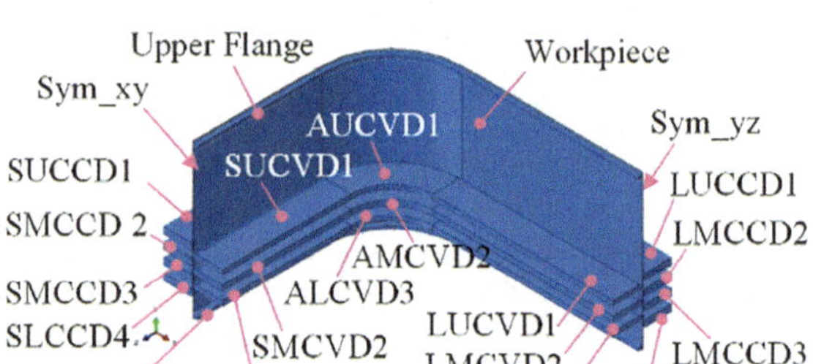

Fig. 3 The FE model of the mechanical bulging process

3 Results and Discussion

3.1 Wave Forming and Equivalent Strain Distribution

The specific waveform forming process can helps the researcher to understand how the material flows, which is very helpful for the improvement of the forming process. The mechanical bulging process of the rectangular bellow and the equivalent strain distribution at the round corner of each layer are shown in Fig. 4. During the forming process of each wave, the deformed parts of the billet were bent and stretched outwards due to the moving convex die, whilst the non-deformed parts remained upright with the help of the fixed dies, avoiding wrinkling on the billet, as shown in Fig. 4a–f. The concave and convex dies were kept still to maintain pressure on the waveforms after the waves were formed. The metal at the straight edges of the bellow flowed mainly in the height direction of the billet, where the elements were mainly bent during the whole process. Nevertheless, the metal at the round corner flowed along the height direction of the billet and the circumferential direction of the round corner, where the elements were greatly stretched and bent.

The strain distribution reflects how the material has been changed. It's beneficial for researchers and manufacturers to judge the degree of metal deformation and identify possible weaknesses. The equivalent strain distribution at the round corner was particularly concerning due to the metal at the round corner was deformed severely. From Fig. 4g–l, the distribution of equivalent strain at the round corner of each layer was similar, and the maximum value of the equivalent strain was 0.406.

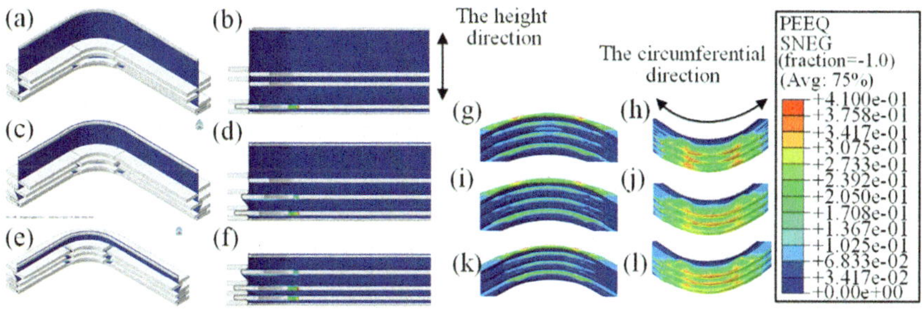

Fig. 4 The forming process of waves and equivalent strain at the round corner: the first wave at the view **a** axonometric and **b** Z direction; the second wave at the view **c** axonometric and **d** Z direction; the third wave at the view **e** axonometric and **f** Z direction; the equivalent strain of layer 1 at **g** inside and **h** outside; the equivalent strain of layer 2 at **i** inside and **j** outside; the equivalent strain of layer 3 at **k** inside and **l** outside

3.2 Dimensions and Wall Thickness of the Formed Bellows

The overall dimensions of the formed rectangular bellow are shown in Fig. 5. The errors of the long and short edges of the rectangle were less than 0.1 and 0.4%, respectively. The inner diameter of the round corner and the height of the rectangular bellow were very close to the designed dimensions, with errors not exceeding 0.11%.

The wall thickness distribution is an important indicator to evaluate the forming quality. Severe wall thickness thinning will weaken the mechanical properties of the rectangular bellow. Serial continuous elements of each layer were selected to measure the wall thickness. As shown in Fig. 6, 9 paths (L1–L9) located at the short edge, the round corner, and the long edge of each layer represented different measuring positions. There were 210 elements on each path, and the first element at the top of each path was numbered 1.

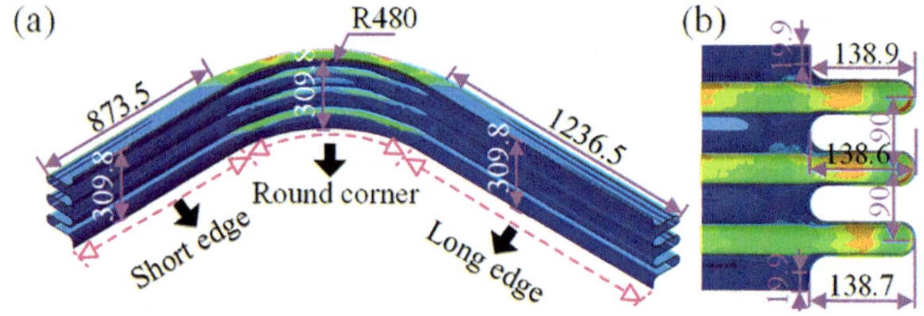

Fig. 5 Basic dimensions of the formed bellows: **a** general; **b** dimensions of waves

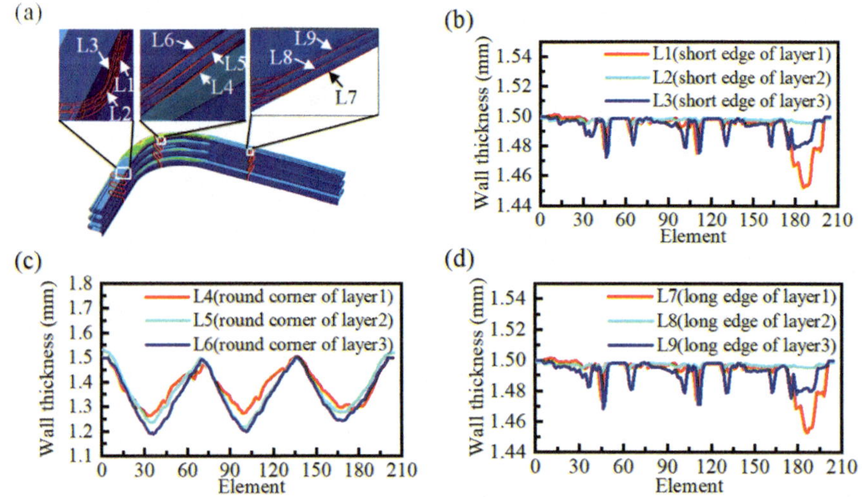

Fig. 6 **a** The measuring positions of the rectangular bellow; the measuring results of the wall thickness at **b** short edge; **c** round corner; **d** long edge

From Fig. 6b–d, the wall thickness distribution on each layer was very similar at the equivalent measuring location. The maximum wall thickness thinning rate of the short straight edge of layers 1–3 were 3.16, 0.47, and 1.82%, respectively, whilst the ones of the long straight edge were 3.12, 0.52, and 2.09%, respectively. The maximum wall thickness thinning rate of layers 1–3 at the round corner were 15.53, 18.74 and 20.48%, respectively, which were greater than the ones at straight edges. Wall thickness is closely related to the metal flowing during the mechanical bulging process. The metal at the straight edges mainly flowed along the height direction. However, the metal at the round corner was stretched significantly along the circumferential direction, resulting in a greater wall thickness reduction at the corner than at the straight edge. Regions with larger wall thickness reduction are likely to be the positions where the metal deforms drastically, and strain may be higher here as well, which is one of the issues will be focused when optimizing the forming process in the future.

4 Conclusion

In this work, a novel mechanical bulging process was proposed to provide a feasible method for fabricating a large rectangular bellow. Numerical simulation was carried out to investigate the forming mechanism of the waves, and the equivalent strain and the wall thickness of the rectangular bellow were discussed.

(1) The elements at the straight edge of the bellow were mainly bent whilst the ones at the round corner were bent and stretched.
(2) The equivalent strain distribution of each layer was similar to each other, and the maximum value of the equivalent strain was 0.406.
(3) The overall dimensions of the rectangular bellow were very close to the designed ones, and the errors were <0.4%.
(4) The wall thickness distribution at each layer of the bellow was similar. The wall thickness thinning rate at the round corner was higher than that at the straight edges, with a maximum value of 20.48%.

References

1. Hwang, Y.-M., et al.: Latest hydroforming technology of metallic tubes and sheets. Metals **11**(9), 1360 (2021)
2. Liu, J., et al.: Numerical investigation of the effect of material properties on forming metal bellows. Int. J. Adv. Manuf. Technol. **116**, 2425–2436 (2021)
3. Gawande, S.H.: A combined numerical and experimental study on metal expansion bellows for STHE. J. Braz. Soc. Mech. Sci. Eng. **40**, 1–14 (2018)
4. Jiang, L., et al.: Influence of process parameters on thinning ratio and fittability of bellows hydroforming. Int. J. Adv. Manuf. Technol. **107**, 3371–3387 (2020)

5. Liu, J., et al.: Deformation behaviors of four-layered U-shaped metallic bellows in hydroforming. Chin. J. Aeronaut. **33**(12), 3479–3494 (2020)
6. Hu, S., et al.: The influence of the MRE forming medium and axial feeding on the forming quality of thin-walled Inconel 718 bellow manufactured by a bulging process: finite element simulation and experimentation. Int. J. Adv. Manuf. Technol. **112**, 387–400 (2021)
7. Zhang, T., et al.: Local Buckling-Induced Forming Method to Produce Metal Bellows. Chinese Journal of Mechanical Engineering. **36**(1), 16 (2023)
8. Yang, J., et al.: Study on the experiment and simulation of titanium alloy bellows via current-assisted forming technology. JOM **70**(7), 1118–1123 (2018)
9. Nishiguchi, I., et al.: On the pressure buckling of rectangular bellows for fusion reactors. Fusion Eng. Des. **41**(1–4), 323–329 (1998)
10. Xia, J.: Structure Design and Strength Analysis of ITER Bellows. Anhui Agricultural University (2020) (in Chinese)
11. Lim, B.H., et al.: Effects of constant engineering and true strain rates on the mechanical behavior of 304 stainless steel. J. Dyn. Behav. Mater. **3**(1), 76–82 (2017)

Effects of Different Parameter Values on the Performance of CPWC-Based Local PWV Estimation

Li Xiong, Yufeng Zhang, Bingbing He, Kexin Zhang, Jingying Zhu, and Xun Lang

Abstract Local pulse wave velocity (PWV) is a widely accepted index for quantitatively evaluating the stiffness of regional arteries, its predictive value is important for assessing the risk of various cardiovascular diseases. High frame rates are required to capture the propagation of pulse wave because it travels at high speed. Plane wave technique, in which all transducer elements are excited at the same time for a full aperture transmission of the region of interest thereby creating a high frame rate, and therefore is suitable for local PWV estimation. However, there have been far less systematic studies focused on the effects of the number and transmission angles of plane waves on the performance of coherent plane wave compounding (CPWC)-based local PWV estimation. In this study, the quantitative effects of these two parameters are assessed by a carotid vessel phantom. First, the blood mimicking fluid is pumped through the phantom via CompuFlow1000. And the linear array probe is attached to the Verasonics system for data acquisition of 3 and 5 plane wave sequences with transmission angles ranging from 1° to 8° and an interval increment step of 1°. The local PWV is then estimated by repeatable and robust ultrasonic transit time method The quantitative effects of different configurations on the performance are assessed by mean values and standard deviations of the estimated local PWVs. The results show that 3 plane waves with transmission angles of (−1°, 1°, 0°) is the best configuration by providing the optimal performance for local PWV estimation. The quantitative findings can provide important advice and improve direction for the CPWC-based local PWV estimation.

Keywords Local pulse wave velocity · Coherent plane wave compounding · Number of plane wave · Transmission angles of plane waves · Optimal parameter values

---

L. Xiong · Y. Zhang (✉) · B. He · J. Zhu · X. Lang
Electronic Engineering Department Information School, Yunnan University, Kunming, China
e-mail: 727786624@qq.com; zhangyf@ynu.edu.cn

K. Zhang
Department of Cardiology, the Second Affiliated Hospital of Kunming Medical University, Kunming, China

W. Wang et al. (eds.), *Computational Statistics and Data Intelligence*, Springer Proceedings in Mathematics & Statistics 463,
https://doi.org/10.1007/978-981-97-4438-1_16

1 Introduction

The stiffness of arteries lead to serious cardiovascular and cerebrovascular diseases, such as hypertension and cerebral embolism [1]. Since the common carotid artery (CCA) is one of the frequent and the earliest sites of arterial stiffness, non-invasive quantitative evaluation of the stiffness of CCA provide positive significance to early diagnosis and effective prevention of cardiovascular and cerebrovascular diseases [2]. Pulse wave velocity (PWV) is a worldwide accepted surrogate index for quantifying the arterial stiffness because it is directly related to the elasticity of arteries [3]. However, the stiffness may take place at any location of the blood vessels and cause vessels minor lesions, which are easily been overlooked. Unlike the global PWV that provides information for the long segment of arteries so that initial changes of arterial stiffness may be concealed, the local PWV provides the result of arterial stiffness close to the tiny site where the illness occurs [4].

Various ultrasound-based techniques, such as flow-area loop method, phased-tracking method and transit time (TT) method [5] have been employed for the local PWV estimation. In particular, the TT method determines the local PWV using linear fitting of the time delay (TD) and the known distance of pulse wave propagation, and thus exhibits more simpleness and robustness on local PWV estimation than other methods.

The local PWV is generally 1–10 m/s, and will be faster with the development of arterial stiffness. High scan line density and high temporal resolution are beneficial to capturing the high speed propagation of pulse wave in a short distance and a short TD. However, traditional focused ultrasound technique uses focused beams to scan region of interest (ROI) line by line, resulting in a relative low frame rate of a few hundred hertz. In plane wave technique, all transducer array transmit full-aperture to the ROI and thus have high frame rate and high scan line density, which opens more possibilities to improve the performance of local PWV estimation based on TT method. However, the plane wave technique is limited by the poor quality of the RF echo signal due to the unfocused transmission, but fortunately, this limitation has been solved by the coherent plane wave compounding (CPWC) technique [6]. The transmission focusing is restored by coherent summation of plane waves with different transmission angles, and thus the penetration is improved effectively. The overall quality of resulting RF echo signals is, therefore, comparable to that of the traditional focused ultrasound technique, and the frame rate remains in the range of several kilohertz.

Several studies, which based on the CPWC technique, have been carried out and continuously developed for local PWV estimation. Nagaoka et al. [7] used 5 plane waves with transmission angles of (−18°, 18°, −9°, 9°, 0°) to visualize the propagation of pulse wave in right CCA during four phases of a cardiac cycle. The result show that the PWVs of four phases are successfully measured by compound imaging, and the measured PWVs of phase II and phase III are similar to previous findings. Apostolakis et al. [8] validated the pulse wave imaging (PWI) technique using CPWC technique with 5 plane waves of (−4°, 4°, −2°, 2°, 0°) in a phantom.

Nauleau et al. [9] used 3 plane waves with angles of (−10°, 0°, 10°) in vitro and 5 plane waves with angles of (−10°, −5°, 0°, 5°, 10°) in vivo for data acquisition to assess the proposed cross-correlation algorithm method and to make better use of PWI spatio-temporal map. In these studies, the CPWC technique is beneficial to enhance the PWI quality or improve the performance of local PWV estimation. However, there is no consensus on the optimal values of the number and transmission angles of plane waves, which are key parameters in CPWC technique that take an important role on the performance of local PWV estimation.

In this study, the quantitative effects of different parameter values for CPWC-based local PWV estimation are assessed on a carotid vessel phantom, which the Young's elastic modulus is known. In the experiments, the blood mimicking fluid was pumped through the vessel phantom with a pumper (CompuFlow 1000), and a linear array connected to the Verasonics system was placed parallel to the phantom surface, and thus, the ultrasonic beams and the direction of pulse wave propagation is vertical. Then, 3 and 5 plane wave sequences with transmission angles ranging from 1° to 8° and an interval increment step of 1° were implemented for RF data acquisition. After beamformed, the CPWC technique was used for coherently compounded off-line. Thereafter, pulse waves at different compounded beam positions were calculated using the one-dimensional (1-D) cross-correlation algorithm. The 20% threshold (20% of the full amplitude of the maximum and minimum of the pulse wave, that also is TH20), which is less interfered by the reflected wave, was selected as the the time fiduciary point (TFP) to calculate the TD of pulse wave propagation. The time-distance relationship was fitted linearly, and the inverse of the fitted line is the estimated local PWV. The mean values and standard deviations (SDs) of 20 estimated PWVs under each configuration were calculated, and the effects of different configurations on the local PWV estimation was quantitative assessed. Finally, the optimal values of number and transmission angles of plane waves for local PWV estimation were determined.

2 Material and Methods

2.1 Hardware

A commercial carotid vessel phantom with known Young's elastic modulus developed by the Shelley Medical Imaging Technologies (Toronto, Canada), a Verasonics Vantage 128 programmable ultrasound research platform developed by Verasonics Inc. (Kirkland, WA, USA) and a linear array transducer (L14-5/38, Ultrasonix Medical Corporation, Richmond, BC, Canada) were used for data acquisition in this study. The carotid vessel phantom is manufactured using a thin-walled silicone carotid model covered with agar-based tissue-mimicking material [10], and with a CCA measured inner diameter of 8 mm and a thickness of 1 mm. The known Young's elastic modulus of the vessel wall is 1.50 ± 2 MPa. As a result, when the blood-mimicking fluid, which developed by Shelley Medical Imaging Technologies (Toronto, Canada) with a density of 1037 kg/m^3, flowing from the inlet to the outlet of the phantom, the ground truth of local PWV for the phantom is 13.42 ± 0.90 m/s according to Moens-Korteweg equation:

$$PWV = \sqrt{\frac{Eh}{2r\rho}} \tag{1}$$

where and are the Young's elastic modulus and density of the blood-mimicking fluid, respectively, is the internal radius of the blood vessel, and is the thickness of the vessel tube wall. All RF data was computed off-line on a PC using MATLAB R2018 b (The MathWorks, Natick, MA, USA).

2.2 CPWC Technique for Local PWV Estimation

The CPWC technique applied plane waves with different numbers and transmission angles to scan a CCA segment, and then the RF echo signals are obtained for calculating the beams in the compounded range. For each scan, the scanning width is and the received RF echo signals are beamformed to beams. Thereafter, the compounded signals are obtained based on the CPWC technique. According to the tangent formula, the compounded length of the posterior wall can be calculated as

$$l = L - 2 \times (d \times \tan(\theta/180 \times \pi)) \tag{2}$$

where is the distance from the outer edge of the posterior vessel wall to the skin's surface, and is the maximum transmission angle of plane waves. Thereafter, the number of beams in the compounded range of the posterior vessel wall, which is used for local PWV estimation, is calculated as follow:

$$n = \frac{l}{L} - N. \tag{3}$$

2.3 Acquisition Sequence

The data acquisition sequences used in this study were implemented by 3 and 5 plane waves with transmission angles ranging from 1° to 8° and an interval increment step of 1°. In this situation, 3 plane wave sequences with transmission angles of (−1°, 1°, 0°), (−2°, 2°, 0°), …, and (−8°, 8°, 0°) and 5 plane wave sequences with transmission angles of (−2°, 2°, −1°, 1°, 0°), (−4°, 4°, −2°, 2°, 0°), …, and (−8°, 8°, −4°, 4°, 0°) were used, respectively. In all cases, alternate polarities were applied to avoid lateral

Table 1 Parameters of the experiment

Type	Parameter	Value
Plane waves	Transducer type	L14-5/38
	Number of elements	128
	The center frequency [MHz]	6.25
	The sample frequency [MHz]	25
	Pulse repetition frequency [kHz]	12
	Ultrasound velocity [m/s]	1540
	Apodization	No

shifts of the moving targets. The pulse repetition frequency for data acquisition was 8 kHz, which ensured that all compounded frame rates were in the range of kilohertz and meet the requirements for PWV estimation. Twenty sets of ultrasonic RF echo signals were collected for each configuration. And Table 1 enlists the typical parameters of system.

2.4 Data Processing

All RF data were processed off-line to estimate the local PWV. The RF echo signals received by the transducer in each configuration were beamformed to obtain the number of beams. The compounded length of the posterior vessel wall was calculated using Eq. (2), and the compounded beam range of the posterior vessel wall was obtained using Eq. (3). The pulse waves in the compounded range for different configurations, which were used for calculating the TD and estimating the local PWV, were estimated by applying the 1-D cross-correlation algorithm on the CPWC-based compounded RF signals. For more accuracy local PWV estimation, a low-pass Savitzky-Golay digital smoother was used for filtering the pulse waves. After that, TH20s were used as TFPs to calculate TDs of pulse wave propagation, and then, TT-based linear regression was performed between the calculated TDs and the known distances of compounded beam positions. Local PWV was estimated as the reciprocal of the linear regression slope at different number and transmission angles of plane waves. Finally, the mean values and SDs of the 20 estimated local PWVs were used to assess the effects of different parameter values on the performance of local PWV estimation and obtain the optimal parameter values.

3 Results

The CPWC-based compounded areas and compounded lengths of posterior vessel wall were calculated after beamforming and coherently compounding processing. Figure 1 shows examples of the individual and compounded results of B-mode ultrasound images, where Figs. 1d1 and d2 are the compounded images of 3 plane waves with transmission angle intervals of 1° and 8°, respectively. The areas between the two yellow dashed lines in these images represent the compounded areas; thus, the posterior vessel walls between the two red dots are the compounded lengths used for local PWV estimation. As expected, the compounded length of the posterior vessel wall decreases with the increase of the maximum transmission angle of plane waves. Obviously, the compounded length of posterior vessel wall in Fig. 1(d2) (corresponding the maximum transmission angle is 8°) is shorter than that in Fig. 1(d1) (corresponding the maximum transmission angle is 1°). In addition, it is not difficult to find that the artifacts are more pronounced in the B-mode images with individual transmission angles but are effectively suppressed in the compounded images. For example, in the compounded images in Fig. 1d1–d2, the artifacts in the lumen and around the compounded wall are significantly reduced such that the walls are clearer than those in their corresponding individual images in Fig. 1a1–c1 and in Fig. 1a2–c2, respectively. Moreover, a large transmission angle induces significant artifacts as the results in Figs. 1d1 and d2. Therefore, it can be reasonably inferred that plane waves with smaller transmission angles are a better choice for local PWV estimation, due to the longer compounded length of posterior vessel wall is theoretically helpful to improve the performance of local PWV estimation [11].

With the estimated TFPs, that are, TH20s in the different pulse waves, the TDs between the adjacent pulse waves were calculated, and the local PWV was obtained using a linear regression fitting applied to the estimated TDs and the known distances of the entire compounded length of the posterior vessel wall. The coefficient correlation was calculated as a performance index because a higher coefficient correlation is

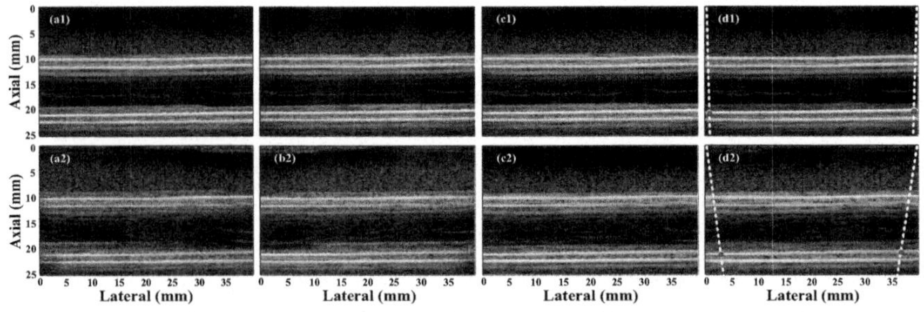

Fig. 1 B-mode ultrasound images of plane waves at 3 different tilted transmission angles and their coherent compounds. **a1–c1** are the images at angles of −1°, 1°, and 0°, and **d1** is their compound; **a2–c2** are the images at angles of −8°, 8°, and 0°, and **d2** is their compound. The areas between the two yellow dashed lines on (**d1**) and (**d2**) represent compounded areas

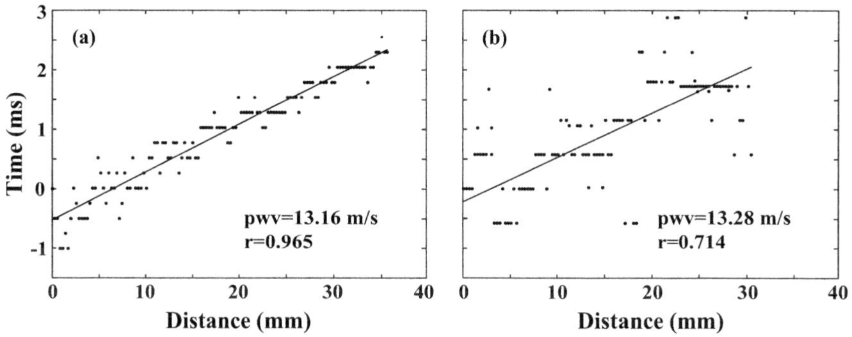

Fig. 2 Dots of beam position (distance) plotted as function of the TD for the PW propagation as well as the linear regression fitted line for the results in Fig. 1

generally associated with a smaller error in linear regression and, thus, with a better performance of the local PWV estimation. As shown in Fig. 2, the dots in the subplots indicate the TDs of the adjacent beam positions, which were calculated according to the TH20s on the filtered pulse waves. The variations in the TDs are in accordance with those of the infer results of Fig. 1. As inferred, the dots in Fig. 2a are closest to the fitting line due to the best precision in the locations of TH20s due to the fewest artifacts and the longest compounded length. The inverse of the slopes of the fitted lines based on the estimated TDs and the known distances of the filtered pulse waves shown in Fig. 2 are the estimated local PWVs: 13.16 m/s (a) and 13.28 m/s (b), which both in the range of ground truth of 13.42° ± 0.90 m/s. As the performance index of linear regression in local PWV estimation, the coefficient correlations of 8° (0.714) is smaller than that of 1° (0.965), which is agreement with the above inference.

To quantitatively assess the effects of different parameters values of performance on the CPWC-based local PWV estimation, the mean values (red solid lines) and SDs (error bars) of 20 estimated local PWVs for all configurations of 3 and 5 plane waves are shown in Fig. 3. The blue, green, and black dashed lines in the Fig. represent the maximum, middle, and minimum ground truth values, respectively. In general, the mean values of the 20 estimated local PWVs are falls within the range of the ground truth of 13.42 ± 0.90 m/s. In addition, the SDs are found to increase with the number and transmission angles of plane waves, which indicates that the accuracy of the local PWV estimation decreased because the artifacts increased. As shown in Fig. 3, for 3(a) and 5(b) plane wave sequences, the transmission angle interval of 1° yields the smallest SDs and provides the best precision. In particular, 3 plane waves with a transmission angle interval of 1° produced the smallest SD, which suggesting 3 plane waves with transmission angles of (−1°, 1°, 0°) is preferred for local PWV estimation.

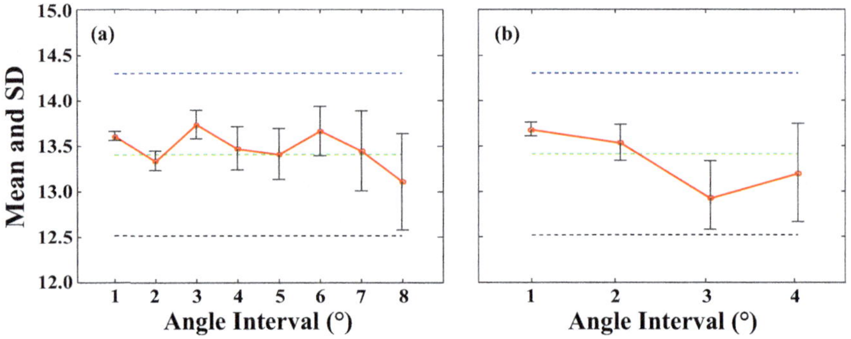

Fig. 3 Mean values (red solid lines) and SDs (error bars) of the estimated local PWVs over 20 acquisitions under the 3(a) and 5(b) plane wave sequences with the different transmission angles. Blue, green, and black dash lines represent the maximum, middle, and minimum values of the true PWV, respectively

4 Conclusion

In this study, the effects of key parameters, i.e., the number and transmission angles of plane waves on the performance of CPWC-based local PWV estimation were assessed by quantitatively analysis, and the optimal parameter values were determined by a carotid vessel phantom with the known Young's elasticity modulus. Different configurations implemented by 3 and 5 tilted plane wave sequences with the transmission angles ranging from 1° to 8° and an interval increment step of 1° were used to acquire 20 sets of RF data for the evaluation experiments. To quantitatively evaluate the effects of parameters on the performance of local PWV estimation, the mean values and SDs of different configurations were calculated. It is found that the mean values are generally in the ground truth, and SD increase with the number and transmission angles of plane waves. The 3 plane waves with transmission angles of (−1°, 1°, 0°) are preferred for CPWC-based local PWV estimation. The effects of these key parameter values obtained in this study can provide important advice and direction for improving the performance of CPWC-based local PWV estimation.

Acknowledgments This work was supported by the Grants (Nos. 62261057 and 202101AS070031) from the National Natural Science Foundation of China, and the Key Project of Fundamental Research of Yunnan Province.

References

1. Izzo, C., Carrizzo, A., Alfano, A., Virtuoso, N., Capunzo, M., Calabrese, M., De Simone, E., Sciarretta, S., Frati, G., Oliveti, M., Damato, A., Ambrosio, M., De Caro, F., Remondelli, P.,

Vecchione, C.: The impact of aging on cardio and cerebrovascular diseases. Int. J. Mol. Sci. **19**(2), 481 (2018)
2. Deng, L., Zhang, Y., Chen, Z., Zhao, Z., Zhang, K., Wu, J.: Regional upstroke tracking for transit time detection to improve the ultrasound-based local PWV estimation in carotid arteries. IEEE Trans. Ultrason. Ferroelectr. Freq. Control **67**(4), 691–702 (2020)
3. Avolio, A.P., Kuznetsova, T., Heyndrickx, G.R., Kerkhof, P.L.M., Li, J.K.: Arterial flow, pulse pressure and pulse wave velocity in men and women at various ages. Adv. Exp. Med. Biol. **1065**, 153–168 (2018)
4. Tang, C.J., Lee, P.Y., Chuang, Y.H., Huang, C.C.: Measurement of local pulse wave velocity for carotid artery by using an ultrasound-based method. Ultrasonics **102**, 106064 (2020)
5. Deng, L., Zhang, Y., Zhao, Z., Zhang, K., Hu, X., Gao, L., Liang, H., Zhang, J.: Ultrasound simulation model incorporating incident and reflected wave propagations along a common carotid artery. Comput. Biol. Med. **104**, 267–277 (2019)
6. Denarie, B., Tangen, T.A., Ekroll, I.K., Rolim, N., Torp, H., Bjåstad, T., Lovstakken, L.: Coherent plane wave compounding for very high frame rate ultrasonography of rapidly moving targets. IEEE Trans. Med. Imaging **32**(7), 1265–1276 (2013)
7. Nagaoka, R., Masuno, G., Kobayashi, K., Yoshizawa, S., Umemura, S., Saijo, Y.: Measurement of regional pulse-wave velocity using spatial compound imaging of the common carotid artery in vivo. Ultrasonics **55**, 92–103 (2015)
8. Apostolakis, I.Z., McGarry, M.D., Bunting, E.A., Konofagou, E.E.: Pulse wave imaging using coherent compounding in a phantom and in vivo. Phys. Med. Biol. **62**(5), 1700–1730 (2017)
9. Nauleau, P., Apostolakis, I., McGarry, M., Konofagou, E.: Cross-correlation analysis of pulse wave propagation in arteries: in vitro validation and in vivo feasibility. Phys. Med. Biol. **63**(11), 115006 (2018)
10. Poepping, T.L., Nikolov, H.N., Thorne, M.L., Holdsworth, D.W.: A thin-walled carotid vessel phantom for Doppler ultrasound flow studies. Ultrasound Med. Biol. **30**(8), 1067–1078 (2004)
11. Huang, C., Ren, T.L., Luo, J.: Effects of parameters on the accuracy and precision of ultrasound-based local pulse wave velocity measurement: a simulation study. IEEE Trans. Ultrason. Ferroelectr. Freq. Control **61**(12), 2001–2018 (2014)

A Robust Localization Algorithm for Intelligent and Connected Vehicles by Variational Bayesian Method

Ziqiang Wang, Ji Mi, Hao Zhu, and Xin Li

Abstract Vehicle localization is a basic and important part of intelligent and connected vehicle technology. However, intelligent and connected vehicles may be affected by outliers when they are driving in complex environments. Additionally, there may be random packet dropouts and delays when positioning data is transmitted by vehicle-to-vehicle communication technology. These factors can cause intelligent and connected vehicles to be unable to obtain real-time and accurate position information, which poses security risks to intelligent transportation. To ensure accurate positioning of intelligent and connected vehicles, this paper takes into account factors such as packet dropouts, delays, and outliers during data transmission. The measurement packet dropouts and randomly delayed data transmissions are described by two Bernoulli variables, while the outliers are modelled as a student's t distribution. As a result, a robust state-space model is developed, and the unknown states and parameters of the proposed model are calculated by the variational Bayesian method. Finally, the experimental results indicate the effectiveness.

Keywords Localization · Variational bayesian · Transmission loss and delay · Heavy-tailed noise · Vehicles

1 Introduction

Intelligent and connected vehicles (ICVs) are crucial to intelligent transportation systems, where accurate vehicle localization is essential [1–3]. ICV positioning technology has two approaches: (1) ICVs can use their own GNSS to improve their positioning accuracy; (2) they can receive positioning data from other ICVs through

Z. Wang (✉) · J. Mi · H. Zhu
School of Automation, Chongqing University of Posts and Telecommunications, Chongqing 400065, China
e-mail: wangziqiang0207@163.com

X. Li
School of Geodesy and Geomatics, Wuhan University, Wuhan 430072, China

W. Wang et al. (eds.), *Computational Statistics and Data Intelligence*, Springer Proceedings in Mathematics & Statistics 463,
https://doi.org/10.1007/978-981-97-4438-1_17

V2V communication to enhance non-self-positioning precision. However, in practice, GNSS signals may be lost or cause multipath problems, leading to measurement outliers and thick-tailed noise [4–7]. Due to factors such as network bandwidth, communication protocols, and sensor failures [6–8], losses and delays are inevitable during data transmission in V2V communication technology.

For the problem of self-location, Havyarimana et al. [9] aim at the problem of ICV driving GNSS signals blocked in urban areas, through the generalized error distribution to approximate non-Gaussian noise and then improving the robustness and accuracy of positioning stability. To meet the security and high-speed performance of ICV in scenarios without GNSS signals, Qin et al. [10] propose a method based on radio frequency identification by reading the tags on the lane and then using V2V communication for information interaction to improve Positioning accuracy. Further, Kaiwartya et al. [11] propose a cyber-physical system without relying on fixed anchors and surrounding equipment positioning information, which converts short and long GNSS signal loss problems into straight lines and curves, respectively. Then, it lowered the effect of positional loss of the GNSS signal. For the non-self-localization problem, Soatti et al. [12] jointly locates static objects around the vehicle and uses them as common noise reference points to improve positioning estimation. Then a distributed localization method is proposed by utilizing V2V communication to share localization and perception information. It is worth noting that the above literature has not considered the impact of communication delay on positioning accuracy. To solve this, Ansari et al. [13] analyze the effect of positioning data delay in V2V communication on the safe driving of ICV and proposed a positioning enhancement system based on cellular communication.

The researchers as mentioned above have not considered the influence of losses and delays in the positioning data in the process of V2V communication and the outlier values of ICV positioning data. Therefore, this paper aims at the problem of ICV non-self-localization and abstracts the problem of estimating the location of other ICVs from the main ICV to a localization problem that simultaneously consider the unknown probability of loss, delay and measurement noise with outliers.

2 Localization Algorithm Based on Variational Bayesian

This paper considers the scenario of two ICVs driving on the road. ICV2 represents the main vehicle, ICV1 represents the vehicle in front, and ICV1 transmits its GNSS positioning data to ICV2 through V2V communication technology. Because the GNSS positioning data may have outliers and induce heavy-tail measurement noise, and the data may have unknown time-varying loss and delay in V2V communication. Therefore, the positioning data received by ICV2 may have unknown time-varying loss, delays and thick-tailed measurement noises. To this end, this paper formulates the ICV localization problem as a state estimation problem with the unknown probability of losses, delays, and measurement noise with outliers:

$$\begin{aligned} x_k &= \boldsymbol{F}_{k-1}x_{k-1} + w_{k-1} \\ z_k &= (1 - \wp_k)\boldsymbol{H}_k x_k + v_k \\ y_k &= \kappa_k z_{k-1} + (1 - \partial_k)z_k, k \geq 2;\, y_1 = z_1 \end{aligned} \tag{1}$$

where k represents time series, x_k is the n-dimension state of ICV1, z_k represents the m-dimension ideal measurement, y_k represents the m-dimension positioning measurement of ICV1 received by ICV2, $\boldsymbol{F}_k$ the state transition matrix, $\boldsymbol{H}_k$ represents measurement matrix, w_k is process noise that obeys Gaussian distribution, and v_k a is heavy-tailed measurement noise.

In this paper, ϑ_k and κ_k follow the Bernoulli distribution and represent the probability of loss and delay, respectively. Thus, we can obtain the following:

$$\begin{aligned} p(y_k|\zeta_k, \chi_k, \chi_{k-1}, \wp_k, \partial_k) = & \left[\mathrm{N}\left(y_k; \boldsymbol{H}_k x_k, \frac{\boldsymbol{R}_k}{\chi_k}\right)\right]^{(1-\wp_k)(1-\partial_k)} \\ & \times \left[\mathrm{N}\left(y_k; \boldsymbol{H}_{k-1}x_{k-1}, \frac{\boldsymbol{R}_{k-1}}{\upsilon_{k-1}}\right)\right]^{(1-\wp_k)\partial_k} \times \left[\mathrm{N}\left(y_k; 0, \frac{\boldsymbol{R}_{k-1}}{\chi_{k-1}}\right)\right]^{\wp_k \partial_k} \\ & \times \left[\mathrm{N}\left(y_k; 0, \frac{\boldsymbol{R}_k}{\chi_k}\right)\right]^{\wp_k(1-\partial_k)} \end{aligned} \tag{2}$$

where $\zeta_k = \left[x_k^{\mathrm{T}}\ x_{k-1}^{\mathrm{T}}\right]^{\mathrm{T}}$ represents the augmented state vector of ICV1, $N(\cdot; \boldsymbol{\mu}, \sum)$ represents the Gaussian PDF, $G(\cdot; \alpha, \beta)$ represents the Gamma PDF, χ_k is an auxiliary parameter and obeys the Gamma distribution.

From (2), the measurement y_k depends on parameters of ζ_k, χ_k, $\wp_k$ and κ_k.

$$\begin{cases} p(\vartheta_k|\tau_k^1) = (\tau_k^1)^{\vartheta_k}(1 - \tau_k^1)^{1-\partial_k} \\ p(\kappa_k|\tau_k^2) = (\tau_k^2)^{\wp_k}(1 - \tau_k^2)^{1-\partial_k} \end{cases} \tag{3}$$

where τ_k^1 and τ_k^2 represent loss and delay respectively and they satisfy $0 \leq \tau_k^1 \leq 1$, $0 \leq \tau_k^2 \leq 1$ and are not correlated with other random variables.

One-step prediction PDF modelling of ζ_k is expressed:

$$p(\zeta_k|y_{1:k-1}) = \mathrm{N}\left(\zeta_k; \hat{\zeta}_{k|k-1}, \boldsymbol{P}_{k|k-1}^{\zeta\zeta}\right) \tag{4}$$

where $\hat{\zeta}_{k|k-1}$ and $\boldsymbol{P}_{k|k-1}^{\zeta\zeta}$ are expressed as follows:

$$\hat{\zeta}_{k|k-1} = \begin{bmatrix} \hat{x}_{k|k-1} \\ \hat{x}_{k-1|k-1} \end{bmatrix}, \boldsymbol{P}_{k|k-1}^{\zeta\zeta} = \begin{bmatrix} \boldsymbol{P}_{k|k-1} & (\boldsymbol{P}_{k-1,k|k-1})^{\mathrm{T}} \\ \boldsymbol{P}_{k-1,k|k-1} & \boldsymbol{P}_{k-1|k-1} \end{bmatrix} \tag{5}$$

where $\hat{x}_{k|k-1}$ $\boldsymbol{P}_{k|k-1}$ can be calculated by Kalman, respectively, and $\boldsymbol{P}_{k-1,k|k-1}$ is given:

$$\boldsymbol{P}_{k-1,k|k-1} = \boldsymbol{P}_{k-1|k-1}\boldsymbol{F}_{k-1}^{\mathrm{T}} \tag{6}$$

The prior PDFs probability of loss and delay can be represented by the Beta distribution:

$$\begin{cases} p\left(\tau_k^1|y_{1:k-1}\right) = Be\left(\tau_k^1; \hat{\alpha}_{k|k-1}^1, \hat{\beta}_{k|k-1}^1\right) \\ p\left(\tau_k^2|y_{1:k-1}\right) = Be\left(\tau_k^2; \hat{\alpha}_{k|k-1}^2, \hat{\beta}_{k|k-1}^2\right) \end{cases} \tag{7}$$

where $\hat{\alpha}_{k|k-1}^1, \hat{\alpha}_{k|k-1}^2, \hat{\beta}_{k|k-1}^1$ and $\hat{\beta}_{k|k-1}^2$ represent the prior shape parameters, which can be expressed as:

$$\begin{cases} \widehat{\alpha}_{k|k-1}^1 = \rho\widehat{\alpha}_{k-1|k-1}^1 \\ \widehat{\beta}_{k|k-1}^1 = \rho\widehat{\beta}_{k-1|k-1}^1 \end{cases} \begin{cases} \widehat{\alpha}_{k|k-1}^2 = \rho\widehat{\alpha}_{k-1|k-1}^2 \\ \widehat{\beta}_{k|k-1}^2 = \rho\widehat{\beta}_{k-1|k-1}^2 \end{cases} \tag{8}$$

where $\widehat{\alpha}_{k-1|k-1}^1, \widehat{\alpha}_{k-1|k-1}^2, \widehat{\beta}_{k-1|k-1}^1$ and $\widehat{\beta}_{k-1|k-1}^2$ the posterior parameter at last moment$k-1$.

Therefore, according to formulas (3) and (7)-(8), the joint conjugate prior can be obtained:

$$\begin{aligned} &p\left(\zeta_k, \chi_k, \chi_{k-1}, \wp_k, \partial_k, \tau_k^1, \tau_k^2|y_{1:k-1}\right) \\ &= \mathrm{N}\left(\zeta_k; \hat{\zeta}_{k|k-1}, P_{k|k-1}^{\zeta\zeta}\right)Be\left(\tau_k^1; \hat{\alpha}_{k|k-1}^1, \hat{\beta}_{k|k-1}^1\right)Be\left(\tau_k^2; \hat{\alpha}_{k|k-1}^2, \hat{\beta}_{k|k-1}^2\right) \\ &\times G\left(\chi_k; \frac{u_k}{2}, \frac{u_k}{2}\right)G\left(\chi_{k-1}; \frac{u_{k-1}}{2}, \frac{u_{k-1}}{2}\right)\left(\tau_k^2\right)^{\wp_k}\left(1-\tau_k^2\right)^{1-\wp_k}\left(\tau_k^1\right)^{\partial_k}\left(1-\tau_k^1\right)^{1-\partial_k} \end{aligned} \tag{9}$$

Let $\Theta_k \triangleq \left\{\zeta_k, \chi_k, \chi_{k-1}, \wp_k, \partial_k, \tau_k^1, \tau_k^2\right\}$, the combined PDF can be described by Bayes' theorem:

$$p(\Theta_k, y_{1:k}) = p(y_k|\zeta_k, \chi_k, \chi_{k-1}, \wp_k, \partial_k)p(\Theta_k|y_{1:k-1})p(y_{1:k-1}) \tag{10}$$

Similar to the solution in [5, 14], each parameter θ in Θ_k can be updated sequentially by variational Bayesian (VB) method. Specifically, let $\theta = \zeta_k$, the results that we can get using variational Bayesian methods can be expressed as:

$$q^{(i+1)}(\zeta_k) \propto \mathrm{N}\left(\bar{y}_k; \bar{H}_k\zeta_k, \tilde{R}_k^{(i+1)}\right)\mathrm{N}\left(\zeta_k; \hat{\zeta}_{k|k-1}, P_{k|k-1}^{\zeta\zeta}\right) \tag{11}$$

where $\overline{y}_k$, $\overline{\boldsymbol{H}}_k$ and $\overline{\boldsymbol{R}}_k^{(i+1)}$ are defined as:

$$\begin{cases} \underset{k}{\bar{y}} = \left[y_k \; y_k\right]^T & \underset{k}{\bar{H}} = diag\left[H_k \; H_{k-1}\right] \\ \tilde{R}_k^{(i+1)} = \left(1/E^{(i)}[1-\vartheta_k]\right) \times \left[\frac{R_k}{E^{(i)}[1-\kappa_k]E^{(i)}[\chi_k]} \; \frac{R_{k-1}}{E^{(i)}[\kappa_k]E^{(i)}[\chi_{k-1}]}\right] \end{cases} \tag{12}$$

Further, $q^{(i+1)}(\zeta_k)$can be approximated, namely:

$$q^{(i+1)}(\zeta_k) = \mathrm{N}\left(\widehat{\zeta}_k; \widehat{\zeta}_{k|k}^{(i+1)}, \boldsymbol{P}_{k|k}^{\zeta\zeta(i+1)}\right) \tag{13}$$

where $\widehat{\zeta}_{k|k}^{(i+1)}$ and $\boldsymbol{P}_{k|k}^{\zeta\zeta(i+1)}$ represent mean and covariance matrices, respectively:

$$\widehat{\zeta}_{k|k}^{(i+1)} = \begin{bmatrix} \widehat{x}_{k|k}^{(i+1)} \\ \widehat{x}_{k-1|k}^{(i+1)} \end{bmatrix}, \boldsymbol{P}_{k|k}^{\zeta\zeta(i+1)} = \begin{bmatrix} \boldsymbol{P}_{k|k}^{(i+1)} & \left(\boldsymbol{P}_{k-1,k|k}^{(i+1)}\right)^{\mathrm{T}} \\ \boldsymbol{P}_{k-1,k|k}^{(i+1)} & \boldsymbol{P}_{k-1|k}^{(i+1)} \end{bmatrix} \tag{14}$$

where the parameters in formula (14) can be calculated by Kalman filter, namely:

$$\begin{cases} \hat{\zeta}_{k|k}^{(i+1)} = \hat{\zeta}_{k|k-1} + K_k^{(i+1)}\left(\underset{k}{\overline{y}} - \widehat{\overline{y}}_{k|k-1}\right) \\ \boldsymbol{P}_{k|k}^{\zeta\zeta(i+1)} = \boldsymbol{P}_{k|k-1} - K_k^{(i+1)}\boldsymbol{P}_{k|k-1}^{\overline{y}\,\overline{y}(i+1)}\left(K_k^{(i+1)}\right)^{\mathrm{T}} \\ K_k^{(i+1)} = \boldsymbol{P}_{k|k-1}^{\zeta\,\overline{y}(i+1)}\left[\boldsymbol{P}_{k|k-1}^{\overline{y}\,\overline{y}(i+1)}\right]^{-1} \\ \boldsymbol{P}_{k|k-1}^{\overline{y}\,\overline{y}(i+1)} = \boldsymbol{S}_k + \tilde{\boldsymbol{R}}_k^{(i+1)} \end{cases} \tag{15}$$

$$\begin{cases} \hat{\zeta}_{k-1|k}^{(i+1)} = \hat{\zeta}_{k-1|k-1} + K_{k-1}^{s(i+1)}\left(\underset{k}{\overline{y}} - \widehat{\overline{y}}_{k|k-1}\right) \\ \boldsymbol{P}_{k-1|k}^{(i+1)} = \boldsymbol{P}_{k-1|k-1} - K_{k-1}^{s(i+1)}\boldsymbol{P}_{k|k-1}^{\overline{y}\,\overline{y}(i+1)}\left(K_{k-1}^{s(i+1)}\right)^{\mathrm{T}} \\ K_{k-1}^{s(i+1)} = \boldsymbol{P}_{k-1,k|k-1}^{\zeta\,\overline{y}(i+1)}\left[\boldsymbol{P}_{k|k-1}^{\overline{y}\,\overline{y}(i+1)}\right]^{-1} \\ \boldsymbol{P}_{k-1,k|k}^{(i+1)} = \boldsymbol{P}_{k,k-1|k-1} - K_{k-1}^{s(i+1)}\boldsymbol{P}_{k|k-1}^{\overline{y}\,\overline{y}(i+1)}\left(K_{k-1}^{s(i+1)}\right)^{\mathrm{T}} \end{cases} \tag{16}$$

$$\begin{cases} \widehat{\overline{y}}_{k|k-1} = \overline{H}_k\hat{\zeta}_{k|k-1} \\ \boldsymbol{S}_k = \overline{H}_k\boldsymbol{P}_{k|k-1}^{\zeta\zeta}\overline{H}_k^{\mathrm{T}} \\ \boldsymbol{P}_{k|k-1}^{\zeta\overline{y}} = \left[\boldsymbol{P}_{k|k-1}, \boldsymbol{P}_{k,k-1|k-1}\right]\overline{H}_k^{\mathrm{T}} \\ \boldsymbol{P}_{k-1,k|k-1}^{\zeta\overline{y}} = \left[\boldsymbol{P}_{k-1,k|k-1}, \boldsymbol{P}_{k-1|k-1}\right]\overline{H}_k^{\mathrm{T}} \end{cases} \tag{17}$$

Similarly, let $\theta = \chi_k$ and $\theta = \chi_{k-1}$, they are updated as Gamma distributions:

$$q^{(i+1)}(\chi_k) = G\left(\chi_k; u_k^{(i+1)}, v_k^{(i+1)}\right), q^{(i+1)}(\chi_{k-1}) = G\left(\chi_{k-1}; g_{k-1}^{(i+1)}, j_{k-1}^{(i+1)}\right) \tag{18}$$

where $c_k^{(i+1)}$ e, $e_{k-1}^{(i+1)}$, $d_k^{(i+1)}$ and $f_{k-1}^{(i+1)}$ can be expressed as:

$$\begin{cases} u_k^{(i+1)} = \frac{m}{2}E^{(i)}[1-\wp_k]E^{(i)}[1-\partial_k] + \frac{m}{2}E^{(i)}[\wp_k]E^{(i)}[1-\partial_k] + \frac{u_k}{2} \\ v_k^{(i+1)} = \frac{1}{2}tr\left(\boldsymbol{D}_k^{(i+1)}\boldsymbol{T}_1/\boldsymbol{A}_k^{(i+1)}\right) + \frac{u_k}{2} + \frac{1}{2}E^{(i)}[\wp_k]E^{(i)}[1-\partial_k]y_k y_k^{\mathrm{T}}/\boldsymbol{R}_k \end{cases} \tag{19}$$

$$\begin{cases} g_{k-1}^{(i+1)} = \frac{m}{2}E^{(i)}[1-\wp_k]E^{(i)}[\partial_k] + \frac{m}{2}E^{(i)}[\wp_k]E^{(i)}[\partial_k] + \frac{u_{k-1}}{2} \\ j_{k-1}^{(i+1)} = \frac{1}{2}tr\left(\boldsymbol{D}_k^{(i+1)}\boldsymbol{T}_2/\boldsymbol{A}_k^{(i+1)}\right) + \frac{u_{k-1}}{2} + \frac{1}{2}E^{(i)}[\wp_k]E^{(i)}[\partial_k]y_k y_k^{\mathrm{T}}/\boldsymbol{R}_{k-1} \end{cases} \tag{20}$$

where $tr()$ is trace function; $\boldsymbol{T}_1$, $\boldsymbol{T}_2$, $\boldsymbol{D}_k^{(i+1)}$ and $\boldsymbol{A}_k^{(i+1)}$ are defined as:

$$\begin{cases} D_k^{(i+1)} = \underset{k}{\bar{H}}\, P_{k|k}^{\zeta\zeta(i+1)}\left(\underset{k}{\bar{H}}\right)^{\mathrm{T}} + \left[\underset{k}{\bar{y}} - \underset{k}{\bar{H}}\,\hat{\zeta}_{k|k}^{(i+1)}\right]\left[\underset{k}{\bar{y}} - \underset{k}{\bar{H}}\,\hat{\zeta}_{k|k}^{(i+1)}\right]^{\mathrm{T}} \\ 0A_k^{(i+1)} = diag\left[\frac{R_k}{E^{(i)}[1-\wp_k]E^{(i)}[\chi_k]E^{(i)}[1-\partial_k]}\ \frac{R_{k-1}}{E^{(i)}[1-\wp_k]E^{(i)}[\chi_{k-1}]E^{(i)}[\partial_k]}\right] \\ \qquad T_1 = diag[I\ 0]T_2 = diag[0\ I] \end{cases} \tag{21}$$

Let $\theta = \vartheta_k, \kappa_k$, we have:

$$\begin{cases} \overset{(i+1)}{\Pr}(\vartheta_k=0,\kappa_k=0) = \Delta^{(i+1)}\exp\Big\{0.5m\mathrm{E}^{(i+1)}[\log\chi_k] + \mathrm{E}^{(i+1)}\left[\log\left(1-\tau_k^1\right)\right] + \mathrm{E}^{(i+1)}\left[\log\left(1-\tau_k^2\right)\right] \\ -0.5\mathrm{tr}\left(\mathbf{D}_k^{(i+1)}\mathbf{T}_1\left(\mathbf{B}_k^{(i+1)}\right)^{-1}\right)\Big\} \\ \overset{(i+1)}{\Pr}(\vartheta_k=0,\kappa_k=1) = \Delta^{(i+1)}\exp\Big\{0.5m\mathrm{E}^{(i+1)}[\log\chi_{k-1}] + \mathrm{E}^{(i+1)}\left[\log\left(1-\tau_k^1\right)\right] + \mathrm{E}^{(i+1)}\left[\log\left(\tau_k^2\right)\right] \\ -0.5\mathrm{tr}\left(\mathbf{D}_k^{(i+1)}\mathbf{T}_2\left(\mathbf{B}_k^{(i+1)}\right)^{-1}\right)\Big\} \\ \overset{(i+1)}{\Pr}(\vartheta_k=1,\kappa_k=0) = \Delta^{(i+1)}\exp\Big\{0.5m\mathrm{E}^{(i+1)}[\log\chi_k] + \mathrm{E}^{(i+1)}\left[\log\left(\tau_k^1\right)\right] + \mathrm{E}^{(i+1)}\left[\log\left(1-\tau_k^2\right)\right] \\ -0.5\mathrm{tr}\left(y_k y_k^{\mathrm{T}}\mathbf{T}_1\left(\mathbf{B}_k^{(i+1)}\right)^{-1}\right)\Big\} \\ \overset{(i+1)}{\Pr}(\vartheta_k=1,\kappa_k=0) = \Delta^{(i+1)}\exp\Big\{0.5m\mathrm{E}^{(i+1)}[\log\chi_k] + \mathrm{E}^{(i+1)}\left[\log\left(\tau_k^1\right)\right] + \mathrm{E}^{(i+1)}\left[\log\left(1-\tau_k^2\right)\right] \\ -0.5\mathrm{tr}\left(y_k y_k^{\mathrm{T}}\mathbf{T}_1\left(\mathbf{B}_k^{(i+1)}\right)^{-1}\right)\Big\} \end{cases} \tag{22}$$

where $\Delta^{(i+1)}$ is a special constant (It's chosen so that the total probability of all possible outcomes or the sum of all estimated values equals one), and the parameter $\boldsymbol{B}_k^{(i+1)}$ is expressed as follows:

$$\boldsymbol{B}_k^{(i+1)} = diag\left[\frac{\boldsymbol{R}_k}{E^{(i+1)}[\chi_k]}\ \frac{\boldsymbol{R}_{k-1}}{E^{(i+1)}[\chi_{k-1}]}\right] \tag{23}$$

Let $\theta = \tau_k^1$ and $\theta = \tau_k^2$, there are:

$$\begin{cases} q^{(i+1)}(\tau_k^1) = Be\left(\tau_k^1; \hat{\alpha}_{k|k}^{1(i+1)}, \hat{\beta}_{k|k}^{1(i+1)}\right) \\ q^{(i+1)}(\tau_k^2) = Be\left(\tau_k^2; \hat{\alpha}_{k|k}^{2(i+1)}, \hat{\beta}_{k|k}^{2(i+1)}\right) \end{cases}, \begin{cases} \hat{\alpha}_{k|k}^{1(i+1)} = \hat{\alpha}_{k|k-1}^1 + E^{(i+1)}[\wp_k] \\ \hat{\beta}_{k|k}^{1(i+1)} = \hat{\beta}_{k|k-1}^1 + E^{(i+1)}[1-\wp_k] \end{cases},$$

$$\begin{cases} \hat{\alpha}_{k|k}^{2(i+1)} = \hat{\alpha}_{k|k-1}^2 + E^{(i+1)}[\partial_k] \\ \hat{\beta}_{k|k}^{2(i+1)} = \hat{\beta}_{k|k-1}^2 + E^{(i+1)}[1-\partial_k] \end{cases} \tag{24}$$

3 Experiments

Assuming that ICV1 and ICV2 are travelling at a constant speed in two-dimensional space, ICV2 obtains the location information of ICV1 through V2V communication and then estimates the location of ICV1. Take into account the discrete-time linear stochastic system represented by the formula (1), where $\boldsymbol{F}_k = \begin{bmatrix} \boldsymbol{I}_2 & \Delta t \boldsymbol{I}_2 \\ 0 & \boldsymbol{I}_2 \end{bmatrix}$, $\boldsymbol{H}_k = \begin{bmatrix} \boldsymbol{I}_2 & 0 \end{bmatrix}$, $\boldsymbol{I}_2$ represent the two-dimensional identity matrix, $\Delta t = 1s$ represents the interval time. The noise w_k, v_k are given:

$$w_k \sim N(0, \boldsymbol{Q}_k), \boldsymbol{Q}_k = \begin{bmatrix} \frac{\Delta t^3 \boldsymbol{I}_2}{3} & \frac{\Delta t^2 \boldsymbol{I}_2}{2} \\ \frac{\Delta t^2 \boldsymbol{I}_2}{2} & \Delta t \boldsymbol{I}_2 \end{bmatrix}, v_k \sim N(0, \Gamma_k), \Gamma_k = \begin{cases} diag\begin{bmatrix} 100 & 10 \end{bmatrix} \text{ w.p.} 1-p \\ 100 \times diag\begin{bmatrix} 100 & 10 \end{bmatrix} \text{ w.p.} p \end{cases} \tag{25}$$

where, $r{=}100m^2$, w.p. indicate "with probability". Initialize parameters $\hat{x}_{0|0} = \begin{bmatrix} 100 & 300 & 100 & 0 \end{bmatrix}^{\mathrm{T}}$ of ICV1 and $\boldsymbol{P}_{0|0} = diag\begin{bmatrix} 100 & 10 & 100 & 10 \end{bmatrix}$. The duration time T = 3000 s, and the number of iterations is $N = 50$ with 200 times independent of Monte Carlo. Initialization parameters $\hat{\alpha}^1_{0|0}, \hat{\beta}^1_{0|0}, \hat{\alpha}^2_{0|0}, \hat{\beta}^2_{0|0}$ to 5, and $c_k, d_k, e_{k-1}, f_{k-1}$ to 1 and δ to 10^{-16}.

Assuming $\rho = 0.99$ and the outlier probability $p = 0.1$ of ICV1. The real probability of loss and delay in V2V communication are generated as follows:

$$p(\tau_k^1) = \begin{cases} 0.10 \le k \le 0.25T \\ 0.30.25T < k \le 0.75T \\ 0.20.75T < k \le T \end{cases}, p(\tau_k^2) = \begin{cases} 0.10 \le k \le 0.25T \\ 0.30.25T < k \le 0.75T \\ 0.20.75T < k \le T \end{cases} \tag{26}$$

Figure 1 shows the RMSE of each algorithm, where RMSE and ARMSE are defined in [5], respectively. The proposed algorithm exhibits superior positioning accuracy compared to RSTKF, as evidenced by the ARMSEpos and ARMSEvel values in Table 1, which are respectively 55.43% and 46.17% lower than those of RSTKF [15]. Additionally, Table I shows that the proposed algorithm has a shorter running time than the existing RSTGAF and a longer single step running time than the existing RSTKF algorithm. In contrast to the rapid peaks and dips of the current RSTKF and RSTGAF algorithms at T = 750 and T = 2250, the proposed algorithm's RMSE curves remain steady, demonstrating its robustness. As a commonly used evaluation metric, a lower RMSE suggests that the proposed algorithm may have improved positioning accuracy and robustness. Finally, Figs. 2 show that the algorithm proposed in this study does a good job of estimating the unknown time-varying losses and delays probability.

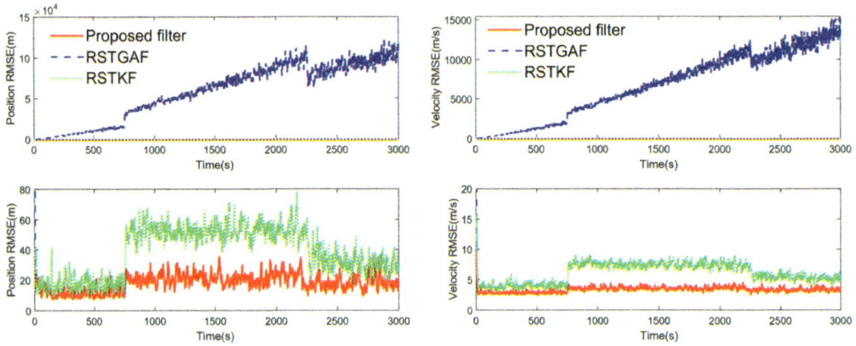

Fig. 1 RMSEpos and RMSEvel of RSTKF, RSTGAF and the proposed algorithm with loss, delay and outliers

Table 1 ARMSEpos, ARMSEvel and single step run time of the RSTKF, RSTGAF and proposed algorithm

Algorithms	ARMS Epos (m)	ARMS Evel (m/s)	Time (ms)
RSTKF [15]	41.91	6.39	1.09
RSTGAF [16]	68,764	8088	7.85
Proposed	18.68	3.44	1.74

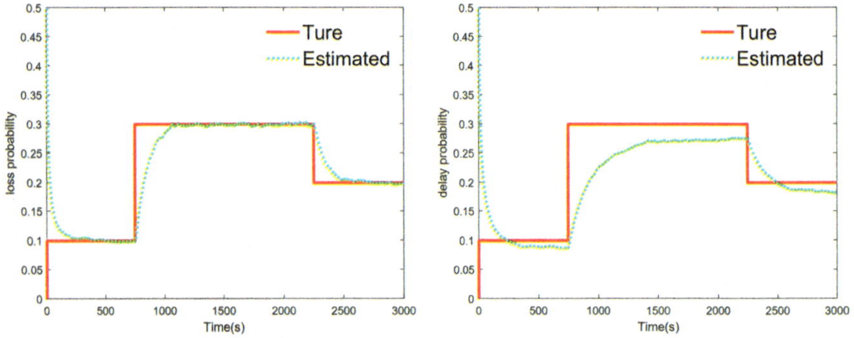

Fig. 2 The true and estimated probability of loss and delay

4 Conclusion

The ICV non-self-localization problem with unknown probability of loss, delay and measurement outliers is abstracted in this paper's proposal for an ICV localization method. We describe the random probability of loss and delay of positioning data using two Bernoulli random variables and simulate measurement outliers of non-self-positioning data. The model with unknown variables is jointly solved by using VB method. Lastly, trials using ICV non-self-positioning data with loss, delay, and outliers simultaneously attest to the usefulness of the positioning method proposed. Collaborative localization to some extent enhances the positioning capability of intelligent connected vehicles. This paper presents a robust localization model for two intelligent connected vehicles. As a next step, this model can be extended to enable co-localization among multiple vehicles.

Acknowledgement The work was supported in part by the Natural Science Foundation of Chongqing under Grant No. cstc2021jcyj-msxmX0373, and by the Scientific and Technological Research Program of Chongqing Municipal Education Commission under Grant No. KJZD-K202200603.

References

1. Zhu, N., Marais, J., et al.: GNSS position integrity in urban environments: A review of literature. IEEE Trans. Intell. Transp. Syst. **19**(9), 2762–2778 (2018)
2. Watta, P., Zhang, X., et al.: Vehicle position and context detection using V2V communication. IEEE Transactions on Intelligent Vehicles **6**(4), 634–648 (2020)
3. Dong, L., Sun, D., et al.: Velocity-free localization of autonomous driverless vehicles in underground intelligent mines. IEEE Trans. Veh. Technol. **69**(9), 9292–9303 (2020)
4. Mahmoud, A., Noureldin, A., et al.: Integrated positioning for connected vehicles. IEEE Trans. Intell. Transp. Syst. **21**(1), 397–409 (2019)
5. Zhu, H., Zhang, G., et al.: An adaptive Kalman filter with inaccurate noise covariances in the presence of outliers. IEEE Trans. Autom. Control **67**(1), 374–381 (2021)
6. Wang, Z., Huang, Y., et al.: An improved Kalman filter with adaptive estimate of latency probability. IEEE Trans. Circuits Syst. II Express Briefs **67**(10), 2259–2263 (2019)
7. Tiwari, R.K., Bhaumik, S., et al.: Particle filter for randomly delayed measurements with unknown latency probability. Sensors **20**(19), 5689 (2020)
8. Zhang, J., You, K., et al.: Bayesian filtering with unknown sensor measurement losses. IEEE Transactions on Control of Network Systems **6**(1), 163–175 (2018)
9. Havyarimana, V., Xiao, Z., et al.: A fusion framework based on sparse Gaussian-Wigner prediction for vehicle localization using GDOP of GPS satellites. IEEE Trans. Intell. Transp. Syst. **21**(2), 680–689 (2019)
10. Qin, H., Peng, Y., et al.: Vehicles on RFID: Error-cognitive vehicle localization in GPS-less environments. IEEE Trans. Veh. Technol. **66**(11), 9943–9957 (2017)
11. Kaiwartya, O., Cao, Y., et al.: Geometry-based localization for GPS outage in vehicular cyber physical systems. IEEE Trans. Veh. Technol. **67**(5), 3800–3812 (2018)
12. Soatti, G., Nicoli, M., et al.: Implicit cooperative positioning in vehicular networks. IEEE Trans. Intell. Transp. Syst. **19**(12), 3964–3980 (2018)
13. Ansari, K.: Cooperative position prediction: Beyond vehicle-to-vehicle relative positioning. IEEE Trans. Intell. Transp. Syst. **21**(3), 1121–1130 (2019)
14. Zhu, H., Wang, Q., et al.: Variational Bayesian Based Localization for Intelligent Vehicle Using Lidar and GPS Data Fusion: Algorithm and Experiments. IEEE/ASME Trans. Mechatron. **27**(6), 5659–5667 (2022)
15. Huang, Y., Zhang, Y., et al.: A novel robust Student's t based Kalman filter. IEEE Trans. Aerosp. Electron. Syst. **53**(3), 1545–1554 (2017)
16. Jia, G., Zhang, Y., et al.: A novel robust Student's t-based Gaussian approximate filter with one-step randomly delayed measurements. Signal Process. **171**, 107496 (2020)

SPRINGER NATURE

GPSR Compliance

The European Union's (EU) General Product Safety Regulation (GPSR) is a set of rules that requires consumer products to be safe and our obligations to ensure this.

If you have any concerns about our products, you can contact us on ProductSafety@springernature.com

In case Publisher is established outside the EU, the EU authorized representative is:

Springer Nature Customer Service Center GmbH
Europaplatz 3
69115 Heidelberg, Germany

The manufacturer's authorised representative in the EU is Springer Nature Customer Service Centre GmbH, Europaplatz 3, 69115 Heidelberg, Germany. If you have any concerns regarding our products, please contact ProductSafety@springernature.com

Printed and bound by CPI Group (UK) Ltd, Croydon, CR0 4YY
01/10/2026
02225919-0016